ANALOG LAYOUT GENERATION FOR PERFORMANCE
AND MANUFACTURABILITY

THE KLUWER INTERNATIONAL SERIES IN ENGINEERING AND COMPUTER SCIENCE

ANALOG CIRCUITS AND SIGNAL PROCESSING
Consulting Editor: **Mohammed Ismail**. *Ohio State University*

Related Titles:

ANALOG LAYOUT GENERATION FOR PERFORMANCE AND MANUFACTURABILITY

by

Koen Lampaert
Katholieke Universiteit Leuven

Georges Gielen
Katholieke Universiteit Leuven

and

Willy Sansen
Katholieke Universiteit Leuven

KLUWER ACADEMIC PUBLISHERS
BOSTON / DORDRECHT / LONDON

A C.I.P. Catalogue record for this book is available from the Library of Congress.

ISBN 978-1-4419-5083-3

Published by Kluwer Academic Publishers,
P.O. Box 17, 3300 AA Dordrecht, The Netherlands.

Sold and distributed in North, Central and South America
by Kluwer Academic Publishers,
101 Philip Drive, Norwell, MA 02061, U.S.A.

In all other countries, sold and distributed
by Kluwer Academic Publishers,
P.O. Box 322, 3300 AH Dordrecht, The Netherlands.

Printed on acid-free paper

Abstract

Analog integrated circuits are very important as interfaces between the digital parts of integrated electronic systems and the outside word. A large portion of the effort involved in designing these circuits is spent in the layout phase. Whereas the physical design of digital circuits is automated to a large extent, the layout of analog circuits is still a manual, time-consuming and error-prone task. This is mainly due to the continuous nature of analog signals, which causes analog circuit performance to be very sensitive to layout parasitics. The parasitic elements associated with interconnect wires cause loading and coupling effects that degrade the frequency behavior and the noise performance of analog circuits. Device mismatch and thermal effects put a fundamental limit on the achievable accuracy of circuits. For successful automation of analog layout, advanced place and route tools that can handle these critical parasitics are required.

In the past, automatic analog layout tools tried to optimize the layout without quantifying the performance degradation introduced by layout parasitics. Therefore, it was not guaranteed that the resulting layout met the specifications and one or more layout iterations could be needed. In this work, we propose a performance driven layout strategy to overcome this problem. In our methodology, the layout tools are driven by performance constraints, such that the final layout, with parasitic effects, still satisfies the specifications of the circuit. The performance degradation associated with an intermediate layout solution is evaluated at runtime using predetermined sensitivities. In contrast with other performance driven layout methodologies, our tools operate directly on the performance constraints, without intermediate parasitic constraint generation step. This approach makes a complete and sensible trade-off between the different layout alternatives possible at runtime and therefore eliminates the possible feedback route between constraint derivation, placement and layout extraction.

Besides its influence on the performance, layout also has a profound impact on the yield and testability of an analog circuit. In this work, we develop a new criterion to quantify the detectability of a fault and combine this with a yield model to evaluate the testability of an integrated circuit layout. We then integrate this technique with our performance driven routing algorithm to produce layouts that have optimal manufacturability while still meeting their performance specifications.

Contents

List of Tables

List of Figures

Chapter 1

Introduction

Our emphasis in this work is on automatic layout generation for analog integrated circuits. However, to gain a global perspective, we first briefly outline the most important steps involved in the design of mixed-signal integrated circuits. Section 1.1 discusses the design cycle of a mixed-signal ASIC. A design methodology for the analog part of the ASIC is presented in section 1.2 and the physical design steps in this flow are discussed in detail in section 1.3. In section 1.4, we will discuss the different layout styles that can be used for analog layout, and we will indicate the scope of the research. An overview of existing tools for analog full-custom layout is given in 1.5, together with a situation of our own work. Finally, in section 1.6, we give a brief overview of the LAYLA tool set, which is the result of this work, and we draw some conclusions in section 1.7.

1.1 Mixed-signal Design Methodology

A mixed-signal ASIC design starts with a high-level specification of the system, follows a series of steps and eventually results in a chip layout. An ideal design flow for a mixed analog/digital chip was proposed in [Donnay 94a, Donnay 94b] and is schematically represented in Fig. 1.1. The design flow consists of the following steps :

- **High-Level Specification** The first step of the design process is to lay down the specifications of the system in a formal, implementation-independent way. To ensure correct functionality of the chip, this high-level specification can be simulated within its system context.

- **High-Level Design** Next, a high-level architecture of the system is designed. During this step, the system is partitioned into different subsystems (digital, DSP and analog) and the high-level specifications are mapped into formal specifications for each subsystem. These specifications are expressed in the language which is most appropriate to describe the type of signals processed by the block (e.g. VHDL for digital, VHDL-A for analog, ...). The correct functionality of the system after high-level design can be verified using mixed-mode simulators.

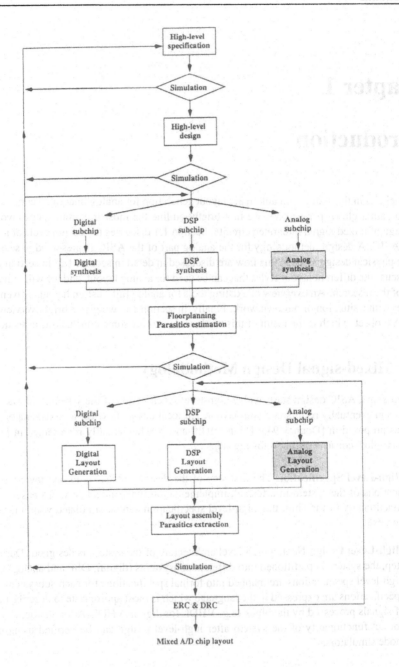

Figure 1.1: The design flow of a mixed A/D chip.

- **Synthesis** Starting from the formal specifications derived during high-level design, each subsystem is designed separately, either by hand or by a dedicated tool set. After this step, the complete floorplan can be generated and the wiring parasitics estimated. The complete chip is then again verified on the transistor level, or if this is impossible, using a mixture of transistor-level and behavioral representations.

- **Layout** After synthesis of the blocks and verification of their functionality, the layout can be generated. The different subsystems are laid out using different layout styles and supporting tool sets. The final chip layout is then assembled using block place and route tools. The real parasitics can now be extracted and included in the netlist for a final verification. If the simulated chip performance is still within the specifications, ERC and DRC can be carried out and the mask production can be started.

This design methodology involves many iterations, both within a step and between different steps. The entire design process may be viewed as transformations of representations in various steps. Each transformation results in a more accurate description of the system which can be analyzed and verified. If a verification shows that the system specifications are violated at a certain point in the design cycle, a number of design steps have to be repeated to correct the error.

The remainder of this work has to do with the physical design aspects of the analog subsystem. Examples of analog subsystems are data acquisition chains and sensor interface circuits. These subsystems also include A/D and D/A converters and analog circuits of which the transfer function is controlled by a digital signal. This means that our analog subsystem will also include some low level digital circuitry such as clocks, control logic and digital registers. Therefore, in the remainder of this work, the analog subsystem will be referred to as "mixed A/D subsystem" or "mixed A/D system".

In the next section, a methodology for the design of such mixed A/D subsystems will be described.

1.2 A Hierarchical Performance-Driven Design Strategy

In order to handle the complexity involved in designing mixed A/D systems, a hierarchical performance-driven design methodology was proposed in [Gielen 92]. The overall design task is broken up into more manageable subtasks, which are again broken up in smaller subtasks, etc. This results in a hierarchy of different levels of design abstraction, where the entities at each level have a different functional abstraction. For mixed-signal systems, the following hierarchical levels can be distinguished (see Fig. 1.2) : mixed A/D (sub)system level (further referred to as system level, e.g. an analog data acquisition chain), module level (ADC, DAC, PLL, ...), circuit level (comparator, opamp, voltage or current reference, ...) and device level (capacitor, resistor, transistor). Each function in the hierarchy can be implemented with different architectures, also called topologies at lower levels. An architecture or topology is defined as an interconnection of blocks from the lower level. The performance-driven design strategy in this hierarchical envi-

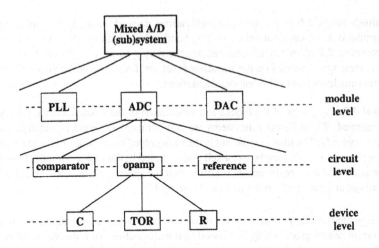

Figure 1.2: Hierarchical levels in mixed A/D design.

ronment is shown schematically in Fig. 1.3. In between any two levels i and $i + 1$ in the design hierarchy, the following steps are performed :

1. top-down

    (a) architecture selection

    (b) sizing and optimization

    (c) verification after sizing

2. bottom-up

    (a) layout generation and extraction

    (b) verification after layout

Conceptually, these steps are the same at any level. First, based upon the required specifications ϕ_i for a block at level i and the technology data and considering criteria such as overall feasibility, smallest area and power consumption, the most appropriate architecture A_i from among a set of alternatives is selected for the block under design. Next, the specifications ϕ_i of the present block are translated into specifications Ψ_j for each of the sub-blocks j within the selected architecture A_i (at the lowest level this is device sizing). The results of the specification mapping process can then be verified by means of mixed-signal behavioral simulation of the complete architecture. If the simulated architecture does not meet its specifications ϕ_i at this point, iterations have to be carried out by redoing some of the previous steps.

This process of architecture selection and specification translation is repeated down the hierarchy until a level is reached which allows a physical implementation B_j (device level for custom

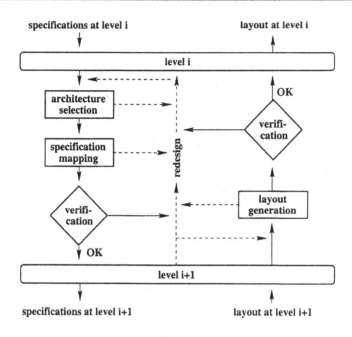

Figure 1.3: The performance-driven analog design strategy.

layouts, higher levels when using standard cells). The hierarchy is then traversed bottom-up, at each level generating the block layout B_i by assembling the sub-block layouts B_j of the architecture. After extraction and detailed verification (by simulation) the layout is passed to the next level up. This results in the layout of the overall system at the top level. A final verification can then be performed by extracting the design from the layout and simulating the whole system with an appropriate mixed-level mixed-signal simulator. If the specifications are not met, redesign iterations have to be carried out again.

This hierarchical performance driven design methodology has been implemented in the AMGIE environment developed in ESAT/MICAS [Gielen 95a]. AMGIE is a CAD system for the automated design of integrated analog modules starting from specifications over structure selection and sizing down to layout. The AMGIE design flow (see Fig. 1.4) is organized hierarchically using the hierarchical levels defined above : a module (e.g. an analog-to-digital converter) is composed of different circuits (e.g. opamps and comparators) which in turn consist of devices. The design flow is based on the top-down synthesis/bottom-up layout generation process described earlier.

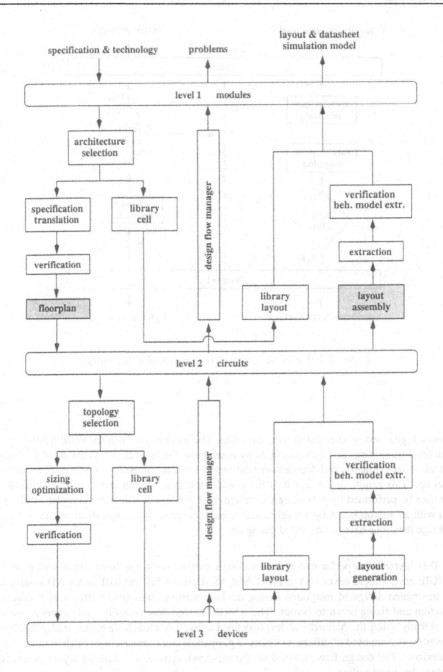

Figure 1.4: The design flow of the analog module generator AMGIE.

AMGIE offers three different design styles to the designer, all handled in the same way by the system : analog standard cells, parameterized cells and full-custom design. The user or the system decides which design style to use. This allows to trade off design flexibility for design speed. The use of predefined standard cells has the advantage that the cells have a proven quality and that the design time is very short. Parameterized cells have a limited set of changeable parameters which offers the additional advantage that the performances can still be tuned over some range. Often, however, the required specifications are different from those of the library cells and optimal fully customized tuning of the circuit is needed. Therefore, full-custom design is necessary and is also offered in AMGIE.

This choice of design styles is also reflected in the design flow of Fig. 1.4. Analog standard cells have a predefined design and layout. Therefore, the specification translation and verification steps can be omitted from the design flow. They also have a predefined layout which is stored in the library. Therefore, the layout generation/assembly steps are not needed and the layout can be taken from the library.

The physical design steps in the design flow (shaded in Fig. 1.4) are the subject of this book and will be discussed in the next section.

1.3 Physical Design Tools for Mixed-signal IC's

1.3.1 Circuit Level Layout Generation

For custom analog designs, a physical implementation becomes possible at the circuit level (opamps, comparators, voltage references ...). At this point in the design flow, the circuit is defined as an interconnection of sized devices : transistors, capacitors and resistors (and spiral inductors for integrated RF circuits). A circuit level layout generator takes as input the specifications and netlist of the circuit as provided by the synthesis tool, together with the constraints on aspect ratio and pin positions, given by the floorplanner or the user. Using this information it generates a fully functional layout for the circuit, which is as compact as possible while satisfying the specifications and the additional geometrical constraints. The most important requirement of an analog circuit level layout generator is to control the influence of all parasitic layout effects such that the performance degradation induced by their combined effects remains within the specifications imposed by the designer or the synthesis system. The most important parasitic effects to be taken into account during circuit level layout generation are interconnect parasitics (including crosstalk), devices mismatches and thermal effects.

1.3.2 System Level Layout Generation

Above the circuit level, the layout generation task has to be divided into two separate tasks. Floorplanning is executed during the top-down part of the design flow, block level layout assembly during the bottom-up part.

1.3.2.1 Floorplanning

After architecture selection, specification mapping and verification, the block under design is defined as an interconnection of blocks from the lower level. At this point in the design flow, the exact area and shape of the sub-blocks are unknown. It is however possible to determine rough estimates of the area and the power consumption of each sub-block since its specifications have been determined. Based on these estimates, the floorplanning algorithm has to determine the position, the aspect ratio and the terminal positions for each sub-block, subject to constraints inherited from the previous floorplanning step and such that the area is minimized and the performance degradation of the circuit is bounded. Parasitic effects to be taken into account include interconnect parasitics, thermal effects and substrate coupling. The aspect ratios and the terminal positions derived by the floorplanner are used as constraints for the floorplanning or the layout generation of the sub-blocks on the next lower hierarchical level. As an additional result of the floorplanning algorithm, a set of estimated values for routing parasitics comes available. These values can be used during synthesis of the lower level blocks.

1.3.2.2 Block Level Layout Assembly

After the physical implementation of the different circuits, a bottom-up place and route step is performed. The task of the block level place and route tool is to repeatedly take the fixed layouts of the different sub-blocks of one hierarchical level and to assemble them into the layout of a block one level higher in the hierarchy until the layout of the complete chip is ready. This requires a block level place and route tool which is capable of controlling performance degradation caused by interconnect parasitics, thermal effects and substrate coupling.

1.3.3 Layout Extraction and Verification

The following tools can be classified under layout extraction and verification :

- **Design Rule Checking (DRC)**
  DRC is executed to verify that the layout meets a set of geometric design rules imposed by the technology process. Design rules specify minimum widths of layout patterns, minimum separations between patterns, etc. DRC must check all patterns in the layout for possible violations of these rules.

- **Extraction**
  An extraction tool regenerates a circuit netlist from the layout of a circuit. The netlist produced by a circuit extractor includes all interconnect parasitics present in the layout and can be used to verify the performance of the circuit after layout.

- **Layout versus Schematic (LVS)**
  An LVS tool compares an extracted circuit netlist with the original circuit netlist to verify its correctness. Differences between the two, such as incorrect connections or device parameters, are flagged to the user.

- **Electrical Rule Checking (ERC)**
  An ERC tool uses the extracted circuit netlist to check the reliability aspects and ensures that the layout will not fail due to electro-migration, self-heat and other effects. ERC tools also check for floating connections and other similar errors.

1.3.4 Scope Of This Work

In this book we will focus on circuit level analog layout generation. The tools have been designed to handle the complexity of typical analog circuits (10-100 devices) and they can control the most important circuit level parasitic layout effects (interconnect parasitics, device mismatch, thermal effects).

Although the tools have been designed for circuit level layout generation, they could also be used for floorplanning and block level layout assembly. The placement tool supports any combination of devices, flexible and fixed blocks, which makes it suitable for the floorplanning and block level layout generation tasks. Most of the parasitics effects encountered on the system level also play a role on the circuit level (interconnect parasitics, thermal effects) and hence are supported by the layout tool. One important exception is the substrate noise problem which becomes more and more important on the system level. Although we have studied the problem and had some ideas for a solution, support for substrate noise problems has not been implemented in the current version of the layout tools.

We have not done any work on layout extraction and verification tools. We are aware of the fact that a lot of problems need to be solved in this area, especially in extraction for deep sub-micron processes. Due to time restrictions, however, these problems have not been studied in this book. Wherever extraction and verification tools were needed in the design flow, existing commercial tools have been used.

1.4 Layout Styles

For layout generation of mixed-signal integrated circuits, several layout styles with different degrees of customization can be used. These approaches mainly differ in density, performance, flexibility and turnaround time, usually trading off one characteristic against another. The layout styles can be broadly classified as either full-custom or semi-custom.

1.4.1 Full-Custom

In a full custom layout style, the layout is done hierarchically in a bottom-up fashion. No restrictions are imposed on the width, height, aspect ratio or terminal positions of the layout blocks at each hierarchical level and they can be placed at any location on the chip surface without any restrictions. Every component in the design is hand crafted for performance, area and power tradeoffs, often resulting in highly irregular placement and routing. This technique has the largest flexibility and the best performance for the highest density since the layout can be tuned and optimized for each application. However, the total turnaround time is quite large if the layout is done by hand and tools to automate this layout technique are very complex.

1.4.2 Semi-Custom

The semi-custom layout styles try to speed up the layout and/or fabrication process by imposing restrictions and constraints on the physical design of the circuits.

- **Standard Cell**

  Standard cell layout architecture considers the layout to consist of rectangular cells of fixed height and variable width. The cells are placed in rows and the space between two rows is called a channel. These channels and the space above and between cells is used to perform interconnections between cells. This layout style introduces several additional constraints on the layout tools. During circuit level layout generation, the circuits have to be laid out with a fixed height. Standard cells usually have their power-supply connections running horizontally on the top and the bottom of the cell, which puts another constraint on the layout process.

- **Mask-Programmable Gate Arrays (MPGA's)**

  In the interest of rapid, low cost manufacturing of small, moderate volume ASICs, some companies offer analog or mixed-signal transistor arrays [THOM 96]. These arrays are developed using a two-step process. In the first step, a base chip is designed with a regular array structure, forming a matrix of predefined transistors and passive components. This design is then mass-produced up through the polysilicon local interconnect stage, with all components and I/O drivers pre-placed (but not connected). Thus, most of the fab process steps are already completed prior to any customer requests for application-specific designs. Once the customer has created a netlist mapped to those predefined components, only the final process steps for metal signal routing remain to complete the design, offering rapid time-to-market potential.

  The use of a gate array has a considerable impact on the layout process. The active and passive components in the array have fixed positions and the placement process is reduced to mapping the circuit devices on the array components. Often, there is no one-to-one correspondence between circuit devices and array components. A circuit resistor for instance can be implemented using a series/parallel combination of array resistors. Due to the restrictions imposed by the fixed array, analog designs laid out in this style can have many layout induced performance problems.

- **Field-Programmable Gate Arrays (FPGA's)**

  The Field-Programmable Gate Array (FPGA) is a new approach to ASIC design that can dramatically reduce manufacturing turn-around time and cost for low volume manufacturing. The architecture of a field-programmable gate array consists of an array of logic blocks. The interconnections between the logic blocks can be programmed to realize different designs. The major difference between FPGA's and MPGA's is that an MPGA is programmed using integrated circuit fabrication to form metal interconnections, while an FPGA is programmed via electrically programmable switches. FPGA's are rarely used for analog designs.

- **Sea of Gates**
  The sea of gates is a semi-custom design style similar to gate array. The master of the sea of gates consists of a predefined regular pattern of logic gates. The interconnect between the predefined gates can be customized by the designer of the chip. Since there are no routing channels on a sea of gates chip, interconnects have to be completed by routing through gates, or by using a second or third layer of metal. The restrictions imposed by the sea of gates styles are similar to those of an array, and analog designs implemented in this style suffer from the same layout induced performance problems. Sea of gates chips are rarely used for analog designs.

1.4.3 Scope Of This Work

In this book we focus on full-custom layout. In normal operating mode, the tools will exploit all degrees of freedom to determine an optimal layout for the circuit subject to the constraints that come from the next higher level in the hierarchy. However, the cost-functions used to drive the place and route tools are general and can be used for semi-custom layout as well. The placement tool can be adapted for semi-custom layout styles by constraining the move-set according to the layout style used (e.g. row based placement for standard cells, a finite number of fixed positions for cells in gate array layout style, etc..).

1.5   Existing Tools for Analog Layout

In order to understand the advantages of the work presented in this book, an overview of existing approaches for analog layout will be given in this section. We will first discuss the analog *macro-cell layout style*, which is used as a basic layout strategy by most automatic analog layout systems. Next, we will discuss a number of macro-cell style analog layout systems, which we will classify based on the way they handle the knowledge required to produce high-quality layouts. This overview by no means intends to be complete. It only provides a representative sampling of existing analog layout programs, in order to depict the state of the art in analog layout generation and to illustrate the variety of ways in which analog constraints and optimizations are handled.

1.5.1   The Analog Macro-Cell Layout Style

In its most general way, the macro-cell layout style follows the steps depicted in Fig. 1.5. *Module recognition*, sometimes also called *circuit partitioning* is the process of dividing the circuit topology into groups of one or more devices, which will be treated as individually placeable and shape-able objects by the placement tool. In the literature, these groups have been called

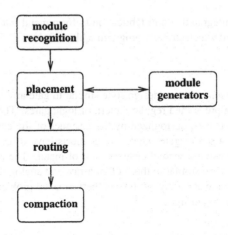

Figure 1.5: The Analog Macro-Cell Layout Style.

macro-cells, structural entities or *modules*. The complexity of these modules ranges from basic devices (transistors, capacitors, resistors) to more complicated structures like current mirrors or differentials pairs. Some analog layout systems restrict the complexity of their modules to basic devices, in which case the module recognition step is trivial and can be omitted. Other layout systems allow more complex modules and rely on the designer or on dedicated algorithms to make the division. An example of a BiCMOS opamp divided into modules is shown in Fig. 1.6.

Once the division into modules is completed, a placement can be generated. Each module can be laid out in several electrically equivalent ways, called *variants*. The task of the *placement program* is to select an optimal variant for each module, and to place these variants on the layout surface in an optimal way. To accomplish this, the placement program uses a set of *module generators* to create a set of possible variants for each module. Typically, these module generators are parameterized programs which procedurally generate layouts for modules based on the module parameters and a number of options. For each module considered, there must be a module generator that generates all the different variants. Maintenance of these module generators over different processes is a hard job. This is also the reason why certain layout synthesis tools try to reduce the number of generators, for instance by restricting the considered modules only to individual devices. A placement of the BiCMOS opamp of Fig. 1.6 is shown in Fig. 1.7.

The next step in the design flow is *routing*. The task of the router is to connect the modules according to the netlist of the circuit. The layout of the BiCMOS opamp after routing is shown in Fig. 1.8. Sometimes, *compaction* is used as a final step to improve the density of the layout.

Although most automatic analog layout systems use a variant of the macro-cell layout style as basic layout strategy, they vary greatly in the way they implement the analog specific features of the various tools. This will be discussed next.

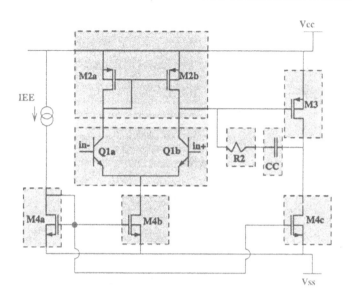

Figure 1.6: BiCMOS opamp divided into modules.

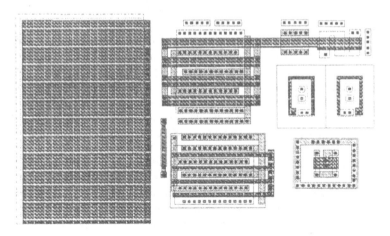

Figure 1.7: Placement for the BiCMOS opamp.

1.5.2 Implementations of the Macro-Cell Layout Style

1.5.2.1 Procedural Tools

The automatic generation of layout through the use of procedural generators [Kuhn 87] is perhaps the most mature layout automation technique used for analog circuits. In these systems, specific

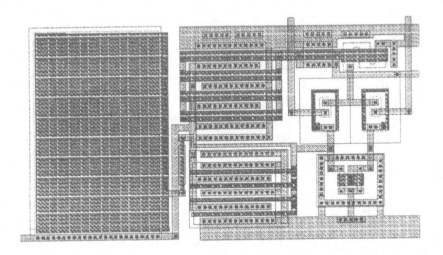

Figure 1.8: Placed and routed BiCMOS opamp.

software is written to support the creation of each unique circuit topology for which layout is desired and the analog-specific knowledge about the layout is coded into the software itself. Parameters passed to the generator at invocation are used to calculate the actual dimensions of the variable devices. A variety of simple constraint mechanisms can be used to adjust the position of the devices to account for their changing sizes. In this way the performance of the generated circuit can be varied within some bounds by changing the invocation parameters to the generator.

Because the topology remains fixed, the base layout can be manually optimized, hence the generated layouts are of reasonably high quality. Virtually all of the commonly used analog-specific layout constraints, e.g. device merging, layout symmetries and matching considerations, can be manually programmed into these generators. The major disadvantage of analog procedural layout generators is that they require a large coding effort for *each* new topology. Because of the relatively high cost of developing new generators, procedural generation is clearly not a good choice for an environment requiring a large variety of circuit topologies. These systems are most useful in environments where many variations of a few general purpose circuit topologies are used, e.g. switched capacitor filter synthesis systems.

1.5.2.2   Template Based

This approach uses a template to capture graphically or textually an expert's knowledge of analog layout for a given circuit type. The template is created once by an expert designer and captures his knowledge of analog specific constraints like symmetry, device matching and parasitic minimization. To generate a circuit layout, one supplies the required electrical parameters for the circuit together with some other geometrical constraints (e.g. circuit shape and aspect ratio). The layout is generated by transforming the template into an actual layout, based on the user-supplied electrical and geometrical parameters.

A typical example of a template driven analog layout system is the "design by example" approach presented by the Philips Research Laboratories [Conway 92]. For a given module type, a sample module layout is created once by an expert designer. This template graphically captures his knowledge of device placement and orientation, routing wire trajectories, material types and widths and position of module terminals. To generate a module, the user supplies the required electrical parameters for each device, the sets of devices which must be matched and geometrical constraints on the module's shape. To reconstruct the placement, the template is analyzed to determine all its possible slicing structures. These structures are stored in an exhaustive slicing tree from which an area-optimal floorplan is derived, depending on specified matching and aspect ratio constraints. The reader is referred to section 4.5.1 for an introduction to slicing trees and their use in analog circuit placement. To obtain real layout, the modules in the template are enlarged and then replaced by the actual module instances, restoring connectivity using a river router [Conway 92]. The final layout is then obtained by compaction.

This technique produces good quality layout in a reasonable amount of time but has some major drawbacks. A template has to be created for each new type of circuit which limits the generality of the method. The library of templates has to be updated for each new technology. For these reasons this method was not considered for this research.

1.5.2.3   Rule Based

Rule-based layout systems offer a fast and flexible integrated set of tools and a rule-based control that can be customized by the designer to meet his specific needs. In these systems, knowledge about analog layout design is incorporated in the rule set. The quality of the final layout depends on the quality of the rule set.

One of the most successful rule-based analog layout systems is ALSYN [Meyer 93]. In the ALSYN design flow, the circuit is first analyzed by a compiled rule set. Using designer specified rules, compound modules are recognized and constraints for placement and routing are determined. ALSYN's placement algorithm is based on the min-cut approach and uses Stockmeyer's algorithm for area optimization. A number of constraints can be specified, e.g. partial slicing specifications, symmetry and clustering groups. Routing can be done either with a grid-based maze routing algorithm or with a grid-less line expansion approach.

At first glance, the rule-based approach seems to be very attractive because every designer can adapt the tools to his specific needs by providing his own set of rules. Unfortunately, this is also the major drawback. The quality of the resulting layout depends for a great deal on the quality of the rule set. Rules are difficult to formulate in a general and context-independent way, and as a consequence, rule-based systems produce acceptable layouts only for the limited range of applications for which the rules have been designed.

1.5.2.4   Algorithmic Approaches

Algorithmic approaches incorporate the analog layout knowledge in the software itself. The main purpose of these algorithmic analog layout tools is to create a layout that is fully functional and optimal in area and performance. They differ from the other described approaches by employing general algorithmic techniques to optimize the layout and to minimize layout induced performance degradation, rather than relying on templates or rule sets.

Most of these tools use cost function driven algorithms for placement and routing. Incremental changes are made to an intermediate solution and the quality of the new solution is evaluated using a cost function. The choice of the cost function is crucial for the operation of the tools. The penalties that are built into this function for violations of analog constraints (matching, crosstalk, loading capacitances, etc.) drive the tools to a solution which minimizes the layout induced performance degradation. The algorithmic layout tools can be classified by the ways of deriving and treating performance constraints.

Heuristic Approaches   First-generation analog layout systems were inspired by algorithms used for digital design that were adapted to satisfy the typical constraints imposed on analog circuit layout, such as device matching, symmetry, and signal decoupling. Layout induced performance degradation was taken into account by classifying nets into categories based on their sensitivity and circuit function, and using weighted penalties in placement and routing cost functions to handle undesired signal coupling and other parasitic layout effects. Although this approach optimized the circuit performance, there was no systematic way of treating performance constraints and deriving circuit sensitivities and there was no guarantee that performance constraints would be met after layout.

One of the first examples of these layout tools was the ILAC program, developed at CSEM [Rijm 88, Rijm 89]. ILAC used a block place and route approach, slicing-tree placement, a floorplan optimization technique based on simulated annealing and routing in channels. Device matching was handled at the device level by using specialized block layout generators (e.g. for current mirrors and matched common centroid bipolar transistors) and at the circuit level by placing blocks in matched groups where all elements have the same orientation and geometrical form. Symmetry was also handled at the device level by using specialized layout generators and at the circuit level by placing blocks in symmetry groups. To avoid coupling between sensitive and noisy nets and to minimize parasitic capacitance and series resistance for sensitive nodes, nets were classified into four categories : sensitive nets such as high-impedance nodes, noisy nets such as output nodes and clock signals, non-critical signal nets and power supply nets. This classification was used to determine routing priorities and to impose penalties for undesired couplings during placement and routing.

A major drawback of ILAC and other similar layout systems is that certain features of the digital layout styles limit the ability to achieve high-quality analog circuit layouts. Some restrictions that help manage the complexity of large digital layouts, e.g. a slicing style placement, and the channel routing style are rather disadvantageous for dense analog circuit-level layouts. In addition, a lot of the analog layout knowledge is built into a large library of device generators, each implementing some common arrangement of basic devices. These libraries are difficult to implement and maintain and a certain degree of technology dependence can not be avoided.

In [Cohn 91] the KOAN/ANAGRAM II tool set was presented as a new layout program that permits more low-level layout optimizations. KOAN is a device-level analog placement tool based on simulated annealing. It differs from other tools in its ability to selectively merge device modules to reduce parasitic capacitance and cell area by appropriate sharing of geometry. KOAN uses a small basic device generator library and creates the more complicated sub-circuit layout structures that make up the bulk of typical generator libraries (e.g. cascode structures, matched differential pairs) dynamically during placement. ANAGRAM II is a detailed general-area router that handles arbitrary grid-less design rules in addition to over-the-device, crosstalk avoiding, mirror-symmetric and self-symmetric wiring. The area-routing strategy incorporates models of capacitive coupling, including simple shielding effects, in its basic evaluation mechanism for paths, allowing path selections to be coerced by possible interactions with other wired nets.

Performance Driven Approaches   Although the tools mentioned in the previous section try to optimize performance in various heuristic ways, none of them *quantifies* performance degradation during layout design. As a consequence, they can not guarantee that the performance specifications will be met after the layout is completed and one or more time-consuming layout-extraction-verification loops may be necessary. An additional problem is that an extracted list of parasitics is usually very large and if performance specifications are not met, no clue is obtained regarding what went wrong and it is not obvious which parasitics have to be changed to reduce the performance degradation.

The solution to this problem is to apply performance driven layout techniques to the analog layout problem. Performance driven layout tools strive to construct a layout such that the performance specifications are guaranteed to be met by construction, by quantifying the layout-induced performance degradation and keeping this below the allowed margins.

The first analog performance driven layout methodology was proposed in [Choudhury 90a, Choudhury 90b]. In this approach, the effect of layout parasitics on circuit performance is modeled using sensitivities. Using this linear approximation and a quadratic optimization technique, the performance constraints for the circuit are mapped to a set of constraints on layout parasitics. These parasitic constraints are then used to drive the layout tools. This methodology has been applied to channel routing [Choudhury 90c], area routing [Malavasi 90, Malavasi 93], placement [Charbon 92, Charbon 94a] and compaction [Felt 93]. Another implementation of a performance driven layout strategy was presented in [Bas 93]. The router presented in this paper uses parasitic constraints to guide the search of a line-expansion area router.

1.5.3   Situation Of This Work

A common feature of all performance driven analog layout tools which have been published up to now is that the performance constraints are mapped to parasitic constraints which are then used to drive the layout tools. In this book, we present an alternative solution to performance driven layout, which eliminates the intermediate constraint generation step, while still guaranteeing a fully functional layout that meets all performance specifications. In our approach, the layout

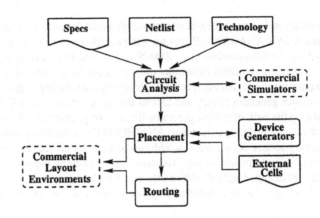

Figure 1.9: Software Architecture of Analog Layout Tool LAYLA

tools are driven *directly* by the performance constraints. We believe that this approach has several advantages which will be explained in chapter 2.

Another limitation of the performance driven layout tools mentioned above is that they focus on the control of performance degradation introduced by interconnect parasitics. The performance driven layout tools presented in this work have been designed to simultaneously handle performance degradation induced by interconnect parasitics, mismatch and thermal effects.

Experiments with our performance driven algorithms show that they find a layout respecting the performance constraints if one exists. Moreover, in all but the most tightly constrained cases, several valid solutions could be found by varying the control parameters of the various algorithms. Although these solutions are equivalent in terms of performance degradation, their quality differs substantially if yield and testability are taken into account. These observations lead us to an extension of the performance-driven routing algorithm to also include yield and testability effects.

1.6   Overview of the Analog Layout Tool LAYLA

Fig. 1.9 gives an overview of a set of tools which have been developed to solve the analog circuit level layout generation problem. Together they provide an integrated framework, called LAYLA, for performance driven analog and mixed-signal layout design. Tools developed within the framework of this research are shown in straight lines, commercial tools that have been integrated in dashed lines.

The input of the tool set consists of three files :

1. **netlist file** The netlist describing the circuit for which the layout has to be designed. This can be a device level netlist, or an architecture of higher level blocks.

2. **specification file** A file with the specifications for the circuit.

3. **technology file** A file describing the process technology.

1.6.1 Circuit Analysis

Before the actual layout process can be started, a number of numerical simulations has to be run to determine the necessary electrical information about the circuit. A circuit analysis program is used to automate this process. The program interfaces to a commercial simulator. It generates a number of input files, runs the simulations and does the post-processing of the simulator output. The most important task of the circuit analyzer is to extract the sensitivities of the circuit performance characteristics with respect to all layout parasitics. The concept of sensitivities and how they are used to calculate performance degradation will be explained in detail in chapter 2.

1.6.2 Device Generation

A library of procedural device generators is used to create the layout for basic devices (transistors, capacitors and resistors). The generators can be called in *interface mode* and in *layout mode*. In interface mode they generate a list of possible layout variants for a device, based on its electrical parameters. Only the information necessary for the placement tool is generated : the bounding box for the device and the terminal geometry. After placement, the device generators are called again in layout mode to generate the complete layout for one selected variant. Special features have been added to the device generators in order to make them suitable for use in a mixed-signal context. The device generators will be discussed in chapter 3.

1.6.3 Placement

The task of the placement tool is to select an optimal position, orientation and implementation for each device in the circuit. The freedom in placing these devices is used to control the layout-induced performance degradation within the margins imposed by the designer's specifications. During each iteration of a simulated annealing optimization algorithm, the layout-induced performance degradation is calculated from the geometrical properties of the intermediate solution. The cost-function is designed to control performance degradation due to interconnect parasitics, mismatch and thermal effects, as will be explained in detail in chapter 4.

1.6.4 Routing

The main task of the performance-driven router is to route the circuit such that the performance degradation caused by the interconnect parasitics remains within the specification margins imposed by the designer. For a given set of circuit specifications, several valid routing solutions can be found. Among these, the routing algorithm selects the solution that additionally maximizes the yield and the testability of the resulting layout. Initially, the circuit is routed with a cost function designed to enforce all performance constraints. After all nets have been routed, the layout parasitics are extracted and the performance of the circuit is verified. In a second phase, nets are ripped up and rerouted to optimize the yield and the testability of the layout. During this process, care is taken not to introduce performance constraint violations. The performance driven routing algorithm together with the yield and testability aspects are the subject of chapter 5.

1.7 Summary and Conclusions

To handle the complexity involved in designing mixed-signal integrated circuits, a hierarchical performance-driven design methodology must be used. The use of automated layout tools within this methodology requires tools that are capable of controlling the various types of layout induced performance degradation that are encountered in mixed-signal chips, at the system level as well as at the circuit level. Existing tools for analog circuit level layout generation can be classified based on the way they incorporate the analog specific knowledge required to control layout-induced performance degradation. In procedural tools, this knowledge is hard-coded into the software. Other layout systems use a template or a rule set to capture an expert's knowledge for a given circuit type. The disadvantages of these approaches is the effort that is required to introduce new circuit topologies. Algorithmic approaches try to overcome this limitation by incorporating analog layout knowledge in the algorithm itself. Most of the existing analog layout programs, however, try to optimize the layout without quantifying the performance degradation. They therefore cannot guarantee that the resulting layout will also meet the specifications. Performance driven layout tools have been proposed to overcome this problem. These tools strive to construct a layout such that the performance specifications are guaranteed to be met by construction, by quantifying the layout-induced performance degradation and keeping this below the allowed margins. LAYLA, the layout program that is described in this book, is an example of a performance driven analog layout system. It implements a direct performance driven layout strategy and simultaneously handles performance degradation induced by interconnect parasitics, device mismatches and thermal effects. In addition to this, it uses the remaining degree of freedom to optimize the manufacturability and testability of the final layout.

Chapter 2

Performance Driven Layout of Analog Integrated Circuits

2.1 Introduction

Generating the layout of high-performance analog circuits is a difficult and time-consuming task which has a considerable impact on circuit performance. The various parasitics which are introduced during the layout phase of an integrated circuit design can introduce intolerable performance degradation. Since these parasitics are unavoidable, the main concern in analog layout synthesis is to control the effects of the parasitics on circuit performance and to make sure that the circuit after layout still performs within its specifications.

Traditionally, this has been done as shown in Fig. 2.1(a). During layout design, the layout is optimized without quantifying the performance degradation. Therefore, it is not guaranteed that the resulting layout will also meet the specifications and a post-layout verification of the circuit with extracted layout parasitics is needed. If it turns out that the circuit does not meet its specifications, one or more time consuming layout iterations are needed. Another problem with this approach is that the extracted list of parasitics is huge, and that the layout designer, or the layout synthesis tool, has no clue which parasitics are responsible for the performance constraint violations.

The goal of a performance driven layout strategy (see Fig. 2.1(b)) is to drive the layout tools directly by performance constraints, such that the final layout, *with parasitic effects* still satisfies the specifications of the circuit.

2.2 Problem Formulation

In this section we will formally describe the performance driven layout generation problem. We will show how a set of performance specifications can be translated into a set of constraints on layout induced performance degradation, with the effect of process variations taken into account.

A *performance characteristic P* is a real number which quantifies some aspect of the performance of a circuit. Examples of performance characteristics are the gain-bandwidth and phase-

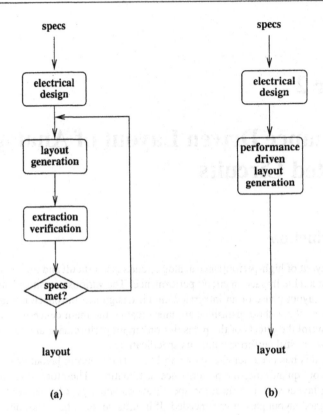

Figure 2.1: Layout methodologies :
       (a) traditional layout methodology,
       (b) performance driven layout.

margin of an opamp or the delay of a comparator. Performance characteristics can be determined using circuit simulation. Depending on the type of analysis that is required to determine their value, performance characteristics are classified as AC, DC or transient characteristics.

A *performance specification* is an interval of the real axis which specifies a set of acceptable values for a performance characteristic P as determined by the application (see Fig. 2.2(a)) :

$$P \in [P_{min}, P_{max}] \tag{2.1}$$

Note that P_{min} can be $-\infty$ and P_{max} can be $+\infty$.

The value of a performance characteristic P is influenced by three kinds of parameters :

1. **Design Parameters**

   Design parameters are the values of device characteristics which can be directly manipulated by the circuit designer, such as the width and the length of a MOSFET transistor, or capacitance and resistance values of passive circuit components.

2. **Process Parameters**
   Process parameters are used to characterize the technology process. Their values are specified by the foundry and can not be controlled by the designer. Examples of process parameters are the K_P and the V_{T0} of a MOS transistor.

3. **Layout Parasitics**
   In general, a layout parasitic can be defined as every cause of performance degradation which is not intended by the circuit designer and whose value is determined by the layout of the circuit. Examples are the parasitic capacitance of a circuit node, or the coupling capacitance between two circuit wires.

Due to parametric fluctuations in the manufacturing process, the process and to some extent also the design parameters are statistical in nature and have to be treated as random variables with a distribution function. In practice, it is impossible to design a circuit in one step, taking all parameters and their stochastic nature into account. Therefore, the design process is usually divided in consecutive steps, during which a subset of the parameters is determined, while the effect of the others is approximated or discarded. The following steps are usually executed :

1. **Electrical Design (Nominal Design)**

   The electrical design of a circuit is the process of determining a set of values for the *design parameters* of a circuit, such that the values of all performance characteristics lie within the specifications of the circuit. During electrical design, the effect of layout parasitics and the stochastic nature of process and design parameters is discarded. Let P_{nom} be the nominal value of performance characteristic P, i.e. the value obtained after electrical design of the circuit (see Fig. 2.2(b)). The actual, measured value of P will differ from its nominal value because of two effects : process variations and layout parasitics. The effect of process variations has to be considered during statistical design, and layout parasitics have to be controlled during layout design.

2. **Yield Estimation (Statistical Design)**

   Due to random variations in the values of the design and process parameters, the actual value of P will deviate from its nominal value. To determine the extent of this deviation, design and process parameters have to be treated as random variables with a distribution function rather than as deterministic variables with a fixed value. As a result, P will also be a random variable with a distribution determined by the circuit and by the distributions of the design and process parameters. A discussion of the techniques which are used to determine the distribution of P, based on the distributions of the different parameters, goes beyond the scope of this book. The reader is referred to the excellent overview articles which exist on this topic [Maly 86, Maly 90, Chang 95]. For the purpose of this discussion, we will assume that the effect of process variations can be represented by an interval

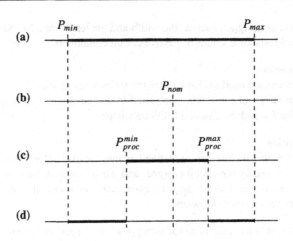

Figure 2.2: (a) performance specification
            (b) nominal performance
            (c) performance variation due to process variations
            (d) performance margins for layout tools.

which contains all possible values of P if process variations are taken into account (see Fig. 2.2(c)) :

$$P \in [P_{proc}^{min}, P_{proc}^{max}] \tag{2.2}$$

3. Layout Generation

The various parasitic layout effects that will be discussed in sections 2.5-2.8 result in an additional performance variation ΔP_{lay}. For the circuit still to be functional, the value of P after layout and with process variations has to be in the interval $[P_{min}, P_{max}]$. It can be derived from Fig. 2.2 that this results in the following two constraints on ΔP_{lay} :

$$P_{proc}^{min} + \Delta P_{lay} > P_{min} \tag{2.3}$$

$$P_{proc}^{max} + \Delta P_{lay} < P_{max} \tag{2.4}$$

From (2.3) and (2.4) we can derive the following constraint for ΔP_{lay} :

$$\Delta P_{lay} \in [\Delta P_{lay}^{min}, \Delta P_{lay}^{max}] \tag{2.5}$$

where

$$\Delta P_{lay}^{min} = P_{min} - P_{proc}^{min} \tag{2.6}$$

and

$$\Delta P_{lay}^{max} = P_{max} - P_{proc}^{max}. \tag{2.7}$$

The goal of a performance driven layout system is thus to generate a layout such that the layout induced performance degradation ΔP_{lay} lies within the interval specified by (2.5), with the limits given by (2.6) and (2.7), for each performance characteristic of the circuit (see Fig. 2.1(b)). Expressions (2.5),(2.6) and (2.7) are the input of a performance driven layout tool.

2.3 Previous Work in Performance Driven Layout Generation

In this section, we give an overview of previous work in performance driven layout generation. The first generation of performance driven layout tools was introduced to control the timing issues associated with the placement of digital circuits. Later on, performance driven techniques were also applied to other physical design problems in digital as well as analog design. We will describe the most important analog performance driven layout tools and conclude this section with a discussion of their shortcomings.

2.3.1 Digital Performance Driven Layout Generation

The performance of digital systems is limited by the delay in propagation of signals through the various paths from the input to the output of the system. In sub-micron technologies, the delay associated with the interconnections may be comparable to or even greater than the delay due to circuit switching [Bak 90]. Therefore, the layout of digital chips has a dramatic impact on the achievable performance and timing constraints have to be taken into account during layout generation. Although this problem is often referred to as *timing driven* layout, it is actually a performance driven layout problem in which the performance constraints are specified as timing requirements. The first performance driven layout algorithms were designed to solve this problem. They can be classified in two major categories, based on the way they handle the timing requirements.

2.3.1.1 Net-Based Approach

In the net-based approach [Dun 84, Ogaw 86, Hau 87, Gao 91], timing requirements are translated into constraints on the length of nets, which are then used to drive the layout algorithms. The place and route tools try to generate a layout such that the constraints on the length of nets are satisfied and hence, also the timing requirements. Several algorithms were proposed to generate the bounds on the net length : examples are the *zero-slack* algorithm in [Hau 87, Nair 89] and a convex programming technique which tries to maximize the flexibility of the layout tools in [Gao 91].

The conversion of timing requirements to net constraints yields a convenient way to perform performance driven layout. However, it also over-constrains the layout tools : the solution space of feasible net lengths satisfying timing constraints is very large and calculating net length bounds before layout is equivalent to picking just one solution. From the viewpoint of the layout algorithms, the chosen net bounds are random. It is not known how to iterate in order to converge to net lengths which would lead to realizable layouts.

2.3.1.2 Path-Based Approach

To overcome these problems, a path-based approach was proposed [Mar 89, Jack 89, Dona 90]. In this approach, the performance of the design is directly optimized during layout. The timing problem is modeled in a path oriented manner, and the performance is dynamically optimized during physical design.

2.3.2 Analog Performance Driven Layout Generation

2.3.2.1 The Berkeley Tools

A similar methodology was later on also adapted to the layout generation for analog circuits. In [Choudhury 90a, Choudhury 90b] an analog constraint-driven layout methodology based on sensitivities was proposed. The goal is to compute the sensitivities of performance functions with respect to all possible layout parasitics before the layout is designed, so that the critical parasitics can be separated and constraints on them can be generated and used to drive the layout tools. Since the performance constraints of a circuit are seen as too abstract for the layout tools to handle directly, a set of constraints on the layout parasitics are derived from these to ensure that the performance constraints are met. If all the parasitic constraints are met by the layout tools, the performance constraints of the circuit should also be met. In general, many possible sets of parasitic constraints that meet the performance constraints can be derived. Using sensitivity information and quadratic optimization, the set of constraints that maximizes the flexibility of the layout tools is computed [Choudhury 90b, Charbon 93]. To achieve this, the constraint generation algorithm reported in [Choudhury 90b, Choudhury 93] and extended in [Charbon 93] used the notion of *layout flexibility* associated with a parasitic constraint to express the ease with which the constraint can be met during layout design. For example, routing a wire with less then $0.1\,pF$ parasitic capacitance is almost impossible and therefore will constrain the routing tool too tightly. Hence, during constraint generation, the sum of the layout flexibilities of all parasitic constraints is maximized, subject to the performance constraints. To model layout flexibility, estimates of maximum and minimum values ($x_{j,min}$ and $x_{j,max}$) for each parasitic x_j are determined and the layout flexibility f_j associated with a parasitic bound $x_{j,bound}$ is modeled with a second-order polynomial (see Fig. 2.3) :

$$f_j = a_j + b_j x_{j,bound} + c_j x_{j,bound}^2 \qquad (2.8)$$

where

$$c_j = \frac{-1}{(x_{j,max} - x_{j,min})^2} \qquad (2.9)$$

$$b_j = -2c_j x_{j,max} \qquad (2.10)$$

$$a_j = -b_j x_{j,min} - c_j x_{j,min}^2. \qquad (2.11)$$

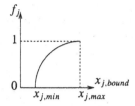

Figure 2.3: Layout flexibility model ([Choudhury 90b])

If the layout tools fail to meet one of the derived parasitic constraints, one or more iterations with another set of constraints is needed. This methodology has been applied for channel routing [Choudhury 90c], area routing [Malavasi 90, Malavasi 93], placement [Charbon 92, Charbon 94a] and compaction [Felt 93].

2.3.2.2 KOAN/ANAGRAM III

In [Bas 93] a technique for *parasitic-bounded routing* was presented. The emphasis in the router is on enforcing hard bounds on parasitics. To achieve this, the search algorithm prunes parasitically non-viable paths as the search progresses. The value of a parasitic is incrementally computed during the path expansion process and if the accumulated value of a parasitic exceeds its bound, the path is no longer considered as a candidate for expansion and is removed from the search tree. The bounds on parasitics can be specified by the designer or generated with a constraint generation algorithm, but no such algorithm was reported in [Bas 93].

2.3.2.3 GELSA

In [Prieto 97] a performance-driven placement algorithm for analog integrated circuits was presented. The placement algorithm is based on simulated annealing and uses a slicing style placement representation (see section 4.5.1). The use of the slicing style placement representation allows to integrate a heuristic global routing algorithm in the placement optimization loop, which results in accurate estimates for interconnect parasitics and routing area. The disadavantage of this approach is that it only works with a slicing style placement representation, which is a poor choice for analog layout, as will be explained in section 4.5.1.

2.3.3 Discussion

A common feature of the presented analog performance driven layout tools is that they are driven by *parasitic constraints*. The performance constraints are seen as too abstract for the tools to handle directly, and are therefore mapped into a set of bounds on parasitics, either by the designer or by a constraint generation algorithm. This approach resembles the net-based approach in digital timing-driven layout, where the high-level timing requirements are mapped into bounds on net parasitics, which are then enforced by the layout tools. Unfortunately, it also suffers from comparable drawbacks.

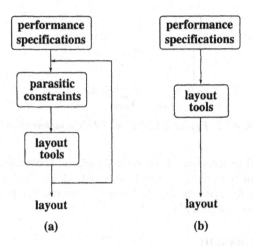

Figure 2.4: Performance driven layout methodologies :
                (a) indirect, with an intermediate constraint generation step,
                (b) direct, without intermediate constraint generation step.

First, converting performance specifications into only *one* set of parasitic bounds is overly constraining. If the tools fail to meet the selected set of parasitic constraints, a new set has to be generated and the layout process has to be repeated (see Fig. 2.4(a)). While this conversion yields a convenient way to handle performance constraints during layout design, it may cause unnecessary iterations. Valid solutions in terms of performance may be rejected because they don't satisfy an a priori selected set of parasitic constraints.

Secondly, the criterion which is used to select a set of parasitics, the flexibility of the layout tools, is something which is very hard, if not impossible to quantify. Although the model (2.8) might be a good approximation in some cases, in general it is far from realistic since it only takes the value of the parasitic constraint into account. The difficulty of embedding a net during routing is also determined by the area of the terminals it has to connect and the distance between them, the obstacles present in the layout area, previously routed nets, etc. Most of these factors are impossible to determine a priori.

It can be concluded that this indirect constraint-based approach overly constrains the layout tools by imposing only one set of parasitic bounds and that the criterion used to determine the set of constraints is of no general use.

2.4   A Direct Performance Driven Layout Strategy

To overcome these drawbacks, we have developed an alternative solution which eliminates the intermediate constraint generation step, while still guaranteeing a fully functional layout that meets all performance specifications. In our approach, the layout tools are driven *directly* by the performance constraints (see Fig. 2.4(b)). The cost functions used to quantify intermediate place and route solutions are based on an evaluation of the performance degradation ΔP_{lay} that would result from accepting the solution : when ΔP_{lay} exceeds the allowed margins, the solution is penalized.

The evaluation of performance degradation is based on a three-step methodology. First, the relevant geometrical information is extracted from an intermediate layout solution. Based on this information, the value of the parameters which model the parasitic effects are calculated. Finally, the influence of the parasitic layout effects on the performance characteristics of the circuit are evaluated using a linear approximation based on performance sensitivities.

This direct method has several advantages. First, by directly taking into account the high-level performance specifications, a complete and sensible trade-off between the different alternative solutions can be made. Secondly, since the performance degradation is calculated at run time, it is not only possible to keep all performance characteristics within their limits, but also to optimize the layout with respect to other constraints, such as yield and testability. Finally, while the methodology described in [Choudhury 93, Charbon 93] can lead to a number of iterations with different parasitic constraints, our tools will either yield a correct layout or will flag the specifications as being impossible to meet, without iterations. This approach therefore eliminates the feedback route between constraint derivation, placement/routing and layout extraction.

2.4.1 Modeling Performance Degradation

Layout induced performance degradation is a complicated and nonlinear function of all parasitic layout effects. However, since the goal of layout design is to limit variations in performance characteristics to be small around their nominal values, several authors [Hong 90, Choudhury 90a, Gyur 90] proposed to use linear approximations using sensitivities to model the dependence of performance on the parasitics.

If the circuit under design has performance characteristics $P_j, j = 1 \ldots N_P$ that are influenced by parasitic layout effects $x_i, i = 1 \ldots N_x$ then the layout induced performance degradation $\Delta P_{j,lay}$ for performance characteristic P_j can be modeled by the following linear approximation :

$$\Delta P_{j,lay} = \sum_{i=1}^{N_x} S_{x_i}^j x_i \qquad (2.12)$$

where $S_{x_i}^j$ is the derivative of performance characteristic P_j to parasitic layout effect x_i :

$$S_{x_i}^j = \frac{\delta P_j}{\delta x_i}. \qquad (2.13)$$

2.4.2 Generation of Performance Sensitivities

Three different methods can be used to compute sensitivities : the *perturbation method*, the *sensitivity network method* and the *adjoint network method*. These three methods can be applied to DC, AC and transient sensitivity computation. Although the details are different, the general principle of the method is the same in the three cases. We will illustrate the different methods for the simplest case, DC sensitivity computation. For details on sensitivity computation, we refer to [Vlach 83, Hoc 85, Dir 69].

The DC solution of a circuit can be computed by solving a system of nonlinear algebraic equations [Vlach 83]:

$$\mathbf{f}(\mathbf{x}^0, \mathbf{h}, \mathbf{w}) = 0 \qquad (2.14)$$

where $\mathbf{x}^0$ is the vector of the voltages and currents, $\mathbf{h}$ is the vector of parameters and $\mathbf{w}$ represents DC sources. The purpose of DC sensitivity analysis is to determine the derivative of one or more elements of $\mathbf{x}^0$ to one or more elements of $\mathbf{h}$.

2.4.2.1 Perturbation Method

In the perturbation method, one parameter is perturbed by Δh, and a new system of the form (2.14) is formulated and solved to determine $\Delta \mathbf{x}^0$. The sensitivity is then approximated by the relation :

$$\frac{\delta \mathbf{x}^0}{\delta h} \approx \frac{\Delta \mathbf{x}^0}{\Delta h} \qquad (2.15)$$

This brute force approach has several disadvantages. First, the finite difference approximation $\frac{\Delta \mathbf{x}^0}{\Delta h}$ tends to the differential sensitivity only in the limit as $\Delta h \rightarrow 0$, and using a very small value for Δh is numerically unstable because of roundoff errors. Second, a complete sensitivity analysis requires the formulation and solution of (2.14) for each component of $\mathbf{h}$ which results in prohibitively high computational cost. The advantage of the approach is its straightforward implementation and general applicability.

2.4.2.2 Direct Method

To evaluate the sensitivity of all components of the vector $\mathbf{x}^0$ to a single parameter h we can differentiate (2.14) with respect to h :

$$\frac{\delta \mathbf{f}}{\delta h} + \frac{\delta \mathbf{f}}{\delta \mathbf{x}}\bigg|_{\mathbf{x}^0} \frac{\delta \mathbf{x}^0}{\delta h} = 0 \qquad (2.16)$$

and rewrite as follows :

$$\mathbf{M}\frac{\delta \mathbf{x}^0}{\delta h} = -\frac{\delta \mathbf{f}}{\delta h} \qquad (2.17)$$

where $\mathbf{M} = \frac{\delta \mathbf{f}}{\delta \mathbf{x}}\big|_{\mathbf{x}^0}$ is the Jacobian matrix about the operating point. To obtain $\frac{\delta \mathbf{x}^0}{\delta h}$ (2.17) can be solved using the LU factors of $\mathbf{M}$, which can be reused from the original Newton-Raphson solution of (2.14). This method generates the sensitivity of the whole vector $\mathbf{x}^0$ with respect to one single variable element h. For each additional parameter h, (2.17) has to be solved again.

2.4.2.3   Adjoint Method

The adjoint method [Dir 69] can be used to compute the sensitivity of any scalar variable $\phi(x^0)$ to a parameter h. To illustrate the method, we restrict $\phi(x^0)$ to be a linear combination of the components of x^0 :

$$\phi = d^T x^0, \tag{2.18}$$

where d is a constant vector. Extension to any function $\phi(x^0, h)$ is possible [Vlach 83]. To compute the sensitivity of ϕ with respect to h, we differentiate (2.18) :

$$\frac{\delta\phi}{\delta h} = d^T \frac{\delta x^0}{\delta h}. \tag{2.19}$$

The formal solution of (2.17) is given by :

$$\frac{\delta x^0}{\delta h} = -M^{-1} \frac{\delta f}{\delta h}. \tag{2.20}$$

Combining equations (2.19) and (2.20) gives :

$$\frac{\delta\phi}{\delta h} = -d^T M^{-1} \frac{\delta f}{\delta h} \tag{2.21}$$

We now define an *adjoint vector* x^a through the relation

$$(x^a)^T = -d^T M^{-1}, \tag{2.22}$$

which can be solved from

$$M^T x^a = -d. \tag{2.23}$$

Substituting (2.22) in (2.21) we get the final form:

$$\frac{\delta\phi}{\delta h} = (x^a)^T \frac{\delta f}{\delta h}. \tag{2.24}$$

For each relevant parameter h, the matrix $\frac{\delta f}{\delta h}$ can be formed and the right-hand side of (2.24) evaluated. As the vectors x^0 and x^a are the same for all parameters h, we see that application of (2.24) requires the solution of only two sets of equations, independent of the number of parameters involved : (2.14) to determine x^0 and (2.23) to determine the adjoint vector x^a. This makes the adjoint method very efficient in situations where the sensitivity of one function of the circuit variables to a set of parameters has to be evaluated.

2.4.2.4 Discussion

To compute performance degradation with the linear approximation (2.12) as proposed in section 2.4.1, the sensitivities of all performance characteristics with respect to all layout parasitics are needed. If there are N_P performance specifications and N_x parasitic layout effects, $N_P \times N_x$ sensitivities need to be computed. From the presentation of the different methods in the previous sections, it can be concluded that the adjoint method of sensitivity computation is best suited for such a calculation. Only two systems of equations have to be solved, irrespective of the number of performance characteristics or parasitics involved.

The problem however is that the adjoint method of sensitivity is not supported in any circuit simulator available for this work. The commercial circuit simulator HSPICE [MS 92] only supports DC sensitivity computation. Sensitivity analysis based on the direct computation method was reported for SPICE3 [Choudhury 88], but our experiments revealed that the implementation was not robust enough for practical use. Therefore, we have implemented the perturbation method in our circuit analysis tool. Although this method suffers from several drawbacks, as discussed in section 2.4.2.2, it offers the advantage of general applicability and easy implementation. An implementation of the direct or adjoint method of sensitivity computation requires access to the source code of a circuit simulator and constitutes a considerable coding effort. Since the focus of this research is on the design and implementation of performance driven layout algorithms, we have used the simpler but less efficient perturbation method. This can simply be substituted by any more efficient sensitivity calculation method when one comes available.

2.4.3 Modeling of Layout Parasitics

In performance driven layout tools, parasitics need to be evaluated frequently and with a reasonable degree of accuracy. The traditional method to evaluate layout parasitics is through the use of extraction tools. Although these tools give very accurate estimates of layout parasitics, their computational requirements prohibit their use in the inner loop of a layout optimization algorithm. In addition to this, extraction tools only work on completed layouts, while performance driven layout tools often need to extract parasitics from incomplete layouts. Placement tools, for instance, have to estimate routing parasitics from an intermediate placement, i.e. prior to actual routing. Therefore, special modeling techniques have to be developed for performance driven layout tools. In the remainder of this chapter, we will discuss the most important layout parasitics and give an overview of their models. Taking into account the requirements discussed above, we will select the most appropriate technique for use during layout optimization.

2.5 Interconnect Parasitics

Interconnect parasitics are a major cause of performance degradation in analog integrated circuits. The non-ideal nature of the wires introduces parasitic elements (capacitors, inductors and resistors) in the circuit, which degrade the performance characteristics. The effect of these parasitic elements is twofold.

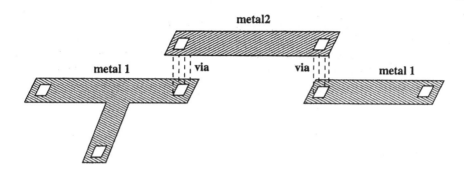

Figure 2.5: Schematic representation of an IC interconnection.

First, they act as additional loads for the devices in the circuit. As such, they cause the poles and the zeros of the circuit to shift, which degrades the small signal characteristics of the circuit. A typical example of this is the parasitic capacitance of the output node of a load-compensated operational transconductance amplifier (OTA), which influences the bandwidth and the phase margin of the circuit. Large-signal characteristics, such as the slew rate, can also be affected.

Secondly, the parasitic elements which are present between wires implementing different nets introduce unexpected signal coupling into a circuit. This effect degrades the noise performance of the circuit or may even destroy its stability through unwanted feedback. Since there is no concept of a noise margin for analog circuits, all noise is generally bad. Many analog signals have very small amplitudes and are thus very sensitive to noise. An example of a performance characteristic which is heavily influenced by parasitic coupling elements is the Power Supply Rejection Ratio (PSRR). A coupling capacitance between one of the input nodes of an amplifier and the power supply node can be disastrous for the PSRR.

2.5.1 Interconnect Modeling

A schematic view of an integrated circuit interconnection is given in Fig. 2.5. The interconnection consists of pieces of conductive material on different layers, connected to each other with vias. To analyze signal propagation in this structure, an equivalent circuit representation has to be constructed, and the values of the elements of the equivalent circuit have to be extracted from the layout. The interconnect modeling problem has to be solved in two steps. The first step is to determine the nature of the equivalent circuit, i.e. what types of circuit elements are important for the considered circuit applications, and the second step is to determine the value associated with each element in the equivalent circuit.

2.5.2 Equivalent Circuit Model

A common starting point for modeling an interconnection is to divide the structure into a number of homogeneous segments, which are each represented by a lumped element equivalent circuit model. The complete interconnection is then modeled as a cascade of lumped element sections.

For an interconnection piece of length D, at a given frequency ω, exact T and Π equivalent circuits can be found. These circuits are shown in Fig. 2.6. In this figure, D is the length of the interconnection segment, Z_0 the characteristic impedance and $\gamma = \alpha + j\beta$ the propagation function. Z_0 and γ are given by :

$$Z_0 = \sqrt{\frac{R_0 + j\omega L_0}{G_0 + j\omega C_0}} \qquad (2.25)$$

$$\gamma = \sqrt{(R_0 + j\omega L_0)(G_0 + j\omega C_0)} \qquad (2.26)$$

where R_0, L_0, G_0 and C_0 are the resistance, inductance, conductance and capacitance per unit length, respectively.

If the length of the interconnection segment is sufficiently small, i.e. $|\gamma D| \ll 1$, the following simplified expressions can be derived for the impedance values of Fig. 2.6 :

$$Z_0 tanh(\frac{\gamma D}{2}) \simeq \frac{(R_0 + j\omega L_0)D}{2} \qquad (2.27)$$

$$Z_0 sinh(\gamma D) \simeq \frac{1}{(G_0 + j\omega C_0)D}, \qquad (2.28)$$

The interconnect model presented above is far too complex for use in a performance driven layout tool and some approximations must be made. For analog circuits operated at relatively low frequency (long wavelength), each wire is much shorter than one wavelength, and can thus be treated as an individual lumped element. A further simplification can be made by observing that $G_0 \ll \omega C_0$ and $\omega L_0 \ll R_0$ is generally true for CMOS analog circuits. (typical values are: $\omega = 2\Pi \times 100 MHz$, $L_0 = 10^{-15}\frac{H}{\mu m}$, $R_0 = 0.01\frac{\Omega}{square}$, $G_0 = 10^{-12}\Omega^{-1}square$, $C_0 = 10^{-16}\frac{F}{\mu m^2}$). This means that the inductance L_0 and the conductance G_0 can be neglected in the models of Fig. 2.6.

Based on the two approximations explained above, we use the interconnect model shown in Fig. 2.7 in our performance driven layout tools. An n-terminal net i is modeled by n parasitic resistors R_{ij}, $j = 1, ..., n$, one parasitic capacitance to ground C_i and a parasitic coupling capacitance C_{ij}, $j = 1, ..., N - 1$ to each other node j in the circuit , where N is the number of nodes in the circuit.

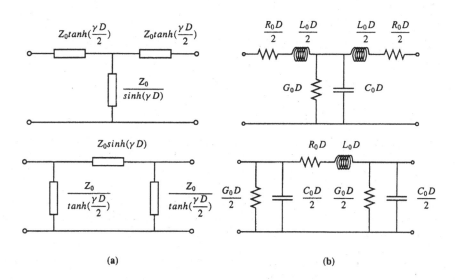

Figure 2.6: *T and Π equivalent circuits for an interconnection of length D*
(a) exact,
(b) approximated.

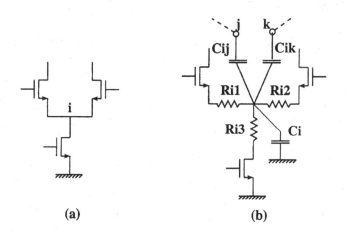

Figure 2.7: *IC interconnection model used in LAYLA.*

2.5.3 Parasitic Extraction

We now explain how the values of the parasitic elements of Fig. 2.7 are calculated.

2.5.3.1 Parasitic Capacitance

Numerical Methods

Exact computation of parasitic capacitances for arbitrary conductor configurations requires the solution of the Laplace equation in two or three dimensions to determine a relation between the conductor voltages and the charge on the conductor surfaces. The solution methods can be classified in two categories. The first category consists of methods that solve the Laplace equation directly: the finite-difference method [Dier 82, Tayl 85, Ueb 86, Seid 88] and the finite-element method [Rueh 73, Ben 76]. Methods of the second category solve the equation in integral form using the boundary-element method [Rao 86, Ning 87]. Multi-pole algorithms have recently been proposed to accelerate boundary-element computations [Nab 91, Nab 92]. These numerical methods give very accurate estimates of the capacitances of complicated three-dimensional structures. Their major drawback is their very high computational cost, which makes them unsuitable for repeated use in automatic layout tools.

Geometrical Methods

Geometrical methods extract interconnect capacitances directly from the layout geometry using parameterized models for commonly encountered interconnect configurations that are stored in a database. Geometrical methods are fast and reasonably accurate, and therefore, they are good candidates for use in a performance driven layout tool.

In our layout tool, we use geometrical methods to compute parasitic interconnect during placement and routing. During placement, calculations must be made based on estimates of the wire topology. During routing, the actual wire topology can be used. Although the details of the method differ for the two problems, the general principle is the same in both cases. The method is based on techniques reported in [Arora 96] and [Choudhury 95].

The layout geometry is first reduced into a set of boxes on different conducting layers. The capacitance of each box is then estimated as the sum of three components (see Fig. 2.8):

- *area capacitance* due to overlap with conductors on different layers. The parasitic capacitance component C_{AXY} can be computed using the following well known equation for parallel plate capacitance :

$$C_{AXY} = \frac{\epsilon}{d} \times A_{XY}, \tag{2.29}$$

where ϵ is the permittivity of the material between the conductors, d is the vertical distance between the conductor layers and A_{XY} is the overlap area.

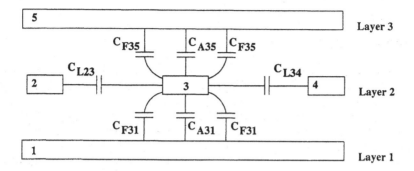

Figure 2.8: Parasitic Capacitance Components :
        area capacitance : C_{AXY},
        lateral capacitance : C_{LXY},
        fringing capacitance : C_{FXY}.

- *lateral capacitance* with respect to conductors on the same layer. For two conductors on the same layer, running in parallel over a length l_{XY} and separated by a distance d, the lateral capacitance component can be computed as :

$$C_{LXY} = F(d) \times l_{XY}, \tag{2.30}$$

where $F(d)$ is the capacitance per unit edge length of two parallel conductors as a function of separation distance d. The function $F(d)$ can be fitted to the following form :

$$F(d) = C_0 + \frac{C_1}{d} + \frac{C_2}{d^2} + \frac{C_3}{d^3} + \frac{C_4}{d^4}, \tag{2.31}$$

where C_0, C_1, C_2, C_3 and C_4 are technology dependent constants that can be determined by fitting (2.31) to simulated or measured capacitance data.

- *fringe capacitance* with respect to conductors on different layers. Fringe capacitance is formed between the edge of one conductor and the surface of a second conductor above or below the first one. The fringe capacitance component can be modeled as :

$$C_{FXY} = C_{F0} \times l_{XY} \times (e^{-\frac{x_2}{x_0}} - e^{-\frac{x_1}{x_0}}) \tag{2.32}$$

where l_{XY} is the parallel length between the conductors, C_{F0} is the maximum value of the fringe capacitance, and x_2 and x_1 are the distances from the edge of the first conductor to the near and far surface edges of the second one. x_0 is a measure of how the fringe capacitance varies for incremental length of the fringing surface. The constants in this model, C_{F0} and x_0 are technology dependent and have to be fitted to simulated or measured data.

2.5.3.2   Parasitic Resistance

Numerical Methods

An accurate prediction of the parasitic resistance of an interconnection again requires the numerical solution of the Laplace equation to determine the relation between the terminal potentials and the terminal currents. The same numerical techniques which are used to compute parasitic capacitance can be used to compute parasitic resistance : the finite-difference method [Harb 86], the finite-element method [Hall 87] and the boundary-element method. These methods yield very accurate results and are very general at the expense of large CPU-times. Therefore they are mostly used in circuit extractors. For the repeated evaluation of parasitic resistance in the inner loop of layout optimization tools a fast and reasonably accurate geometrical method has to be used.

Geometrical Methods

One heuristic technique to find the parasitic resistance of an interconnection is based on a decomposition of the interconnection into rectangles along equipotential lines [Hor 83]. The resistance R_i of a rectangle i can then easily be calculated from its length-to-width ratio :

$$R_i = \rho_{\square.i} \frac{l_i}{w_i} \tag{2.33}$$

where l_i and w_i are the length and the width of the rectangle and $\rho_{\square.i}$ its sheet resistance. The parasitic resistances of an interconnection can then be determined by series/parallel combination of all the R_i.

2.6   Device Parasitics

With the ever decreasing size of integrated circuit devices, the influence of the parasitic elements associated with the layout of devices, such as series resistances and capacitances becomes more and more noticeable. In CMOS circuits, the dominant layout capacitance is generally associated with MOS gate structures. The gate areas, and thus the gate capacitances, are fixed by the designer and cannot be minimized in layout. There are, however, device capacitances which can be controlled by proper layout. For example, the *pn* junctions which form the MOS device source and drain regions each have a non-linear voltage dependent capacitance which consists of two terms, proportional to the junction area and perimeter, respectively :

$$C_{jSBt} = \frac{A_S C_j}{(1 - \frac{V_{BS}}{\phi_j})^{m_j}} + \frac{P_S C_{jsw}}{(1 - \frac{V_{BS}}{\phi_j})^{m_{jsw}}}, \tag{2.34}$$

where, C_{jSBt} refers to the total source (or drain) bulk capacitance, A_S and P_S to the source (or drain) area and perimeter, C_j and C_{jsw} to the bottom and sidewall junction capacitances in absence of any junction voltage and ϕ_j to the built-in junction potential. m_j and m_{jsw} depend on

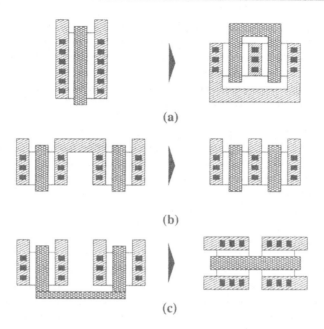

Figure 2.9: Diffusion sharing in MOS transistor layout structures :
(a) folding,
(b) diffusion merging,
(c) connection by abutment.

the doping profile of the junction. Equation (2.34) reveals that this capacitance can be reduced by minimizing the size of all diffusions. In particular, large FET devices can be *folded* to allow a single source or drain diffusion to be shared by two gate regions. Device folding is illustrated in Fig. 2.9(a). An additional large saving in diffusion capacitance can be made by *device merging*, i.e. placing devices such that diffusion geometry is shared between electrically connected devices as shown in Fig. 2.9(b). This type of *geometry sharing* has the additional benefit of improving the layout density. If spacing rules permit, additional capacitance and resistance can be saved by making the connection between some adjacent devices by *abutment*, rather than by explicit wiring (see Fig. 2.9(c)).

2.7 Mismatch

Mismatch is defined in [Pel 89] as *the process that causes process-induced, time-independent random variations in physical quantities of identically designed devices.* Since the functionality of analog circuits is often based on ratios of devices, mismatch puts a fundamental limit on the achievable circuit performance in a particular technology process. The accuracy of D/A converters that rely on static mirrors for example is limited by the achievable transistor matching [Bast 96a]. In switched capacitor filters, the transfer characteristic is determined by the ratios of capacitors. Accurate transistor matching is also required for minimal offset in operational amplifiers. In [King 96], it is shown that under optimal biasing conditions, the performance ratio $Speed.Accuracy^2/Power$ of elementary voltage and current building blocks is fixed and inversely proportional to the technology mismatching.

A good understanding of the matching behavior of components in a particular technology process is thus critical in designing analog ICs and a lot of effort has been spent in developing accurate mismatch models [Shyu 84, Shyu 86, Pel 89, Mich 92, Bast 95]. In addition to this, a number of layout rules are generally accepted for optimum matching, based on assumptions of process gradients, temperature gradients, anisotropic and boundary effects [Vitt 85]. In the remainder of this section, an overview of this work will be given.

2.7.1   Mismatch Model

Mismatch between the parameters of a group of equally designed devices is the result of several random processes which occur during every fabrication phase of the devices. In [Pel 89] physical mismatch causes are divided into two classes of mismatch-generating processes. The first class of mismatch-generating process on a parameter P is spatial "white noise" or short-distance variations, with the following features :

- the total mismatch of parameter P is composed of many single events of the mismatch-generating process;

- the effects on the parameter are so small that the contributions to the parameter can be summed;

- the events have a correlation distance much smaller than the device dimensions.

Consequently, the values of the mismatch of parameter P are normally distributed with zero mean. Examples of mismatch processes of this class are : distribution of ion-implanted, diffused, or substrate ions, local mobility fluctuations, oxide charges, etc. On the other hand, the circular parameter-value distributions which originate from wafer fabrication and the oxidation process are explained by a second class of mismatch : deterministic processes (e.g. gradients) which can be modeled as an additional stochastic process with a long correlation distance.

A mathematical treatment of these two classes of mismatch behavior yields the following model for the mismatch of a parameter P [Pel 89]:

$$\sigma^2(\Delta P) = \frac{A_P^2}{WL} + S_P^2 D^2 \qquad (2.35)$$

where A_P and S_P are technology dependent area and distance proportionality constants for parameter P. Applied to the threshold voltage V_{T0}, the current factor β and the substrate factor K of a MOS transistor, this gives :

$$\sigma^2(V_{T0}) = \frac{A_{V_{T0}}^2}{WL} + S_{V_{T0}}^2 D^2 \tag{2.36}$$

$$\sigma^2(K) = \frac{A_K^2}{WL} + S_K^2 D^2 \tag{2.37}$$

$$\frac{\sigma^2(\beta)}{\beta^2} = \frac{A_W^2}{W^2 L} + \frac{A_L^2}{WL^2} + \frac{A_\mu^2}{WL} + \frac{A_{C_{ox}}^2}{WL} + S_\beta^2 D^2$$

$$\simeq \frac{A_\beta^2}{WL} + S_\beta^2 D^2 \tag{2.38}$$

where A_X and S_X are process-related area and distance constants for parameter X. The validity of this model has been verified by measurements.

In [Bast 95] the mismatch of small size MOS transistors was characterized. The measurements of the current factor mismatch was in good agreement with the model (2.38), but for the area dependence of the threshold voltage mismatch significant deviations from model (2.36) were observed. It was found that the linear dependency of the threshold voltage mismatch on the inverse of the square root of the effective channel area no longer holds for small length transistors. This observation was explained by the fact that for small geometry MOS devices, the channel depletion thickness can no longer be considered uniform: it is a function of the channel length, the channel width and the drain voltage. An extended threshold voltage mismatch model taking these effects into account was given :

$$\sigma^2(V_{T0}) = \frac{A_{1VT}^2}{WL} + \frac{A_{2VT}^2}{WL^2} - \frac{A_{3VT}^2}{W^2 L} + S_\beta^2 D^2 \tag{2.39}$$

This new model is able to accurately predict the threshold voltage mismatch for small size MOS transistors.

2.7.2 Layout Rules for Optimum Matching

It can be concluded from the mismatch models presented in the previous section that the area and the distance between matching devices should be selected based on the required matching degree. If the area of and the distance between two devices are known, equations (2.37), (2.38), (2.39) can be used to predict the mismatch. This value has to be seen as a minimum value of the achievable accuracy. To make sure that the actual mismatch is as close as possible to this lower limit, a number of layout rules for optimum matching have to be followed. In the remainder of this section we will give an overview of the most important layout rules for optimum matching. These rules were collected from various sources. In [Vitt 85], a number of rules were formulated, based on assumptions of process, temperature and stress gradients, anisotropic effects and boundary effects. In [Bas 96b] the influence of different layout styles on MOS transistor matching has been investigated. [McNu 94] discusses various systematic capacitance matching errors and corrective layout procedures. Additional information can also be found in [Laker-Sansen 94].

- **Same Structure**

  Matching devices should have the same structure. For instance, a poly-poly capacitor can not be matched with a metal-poly capacitor. Due to the large spreading on the absolute values of process parameters, they can never be used to design predictable device parameter ratios.

- **Same Temperature**

  Since device parameters are sensitive to temperature variations, matching devices should have the same local temperature. Power dissipating devices cause a temperature distribution across the layout area. Matching devices should be placed on isotherms. The influence of temperature effects on circuit performance will be discussed in section 2.8.

- **Same Shape and Size**

  Matching devices should have equal shapes and sizes. Matching transistor pairs should have the same number of fingers. For matching resistors, not only the number of squares should be equal but also the length, the width and the number of bends (see Fig. 2.10).

- **Common-centroid geometries**

  Four different layout styles for a MOS transistor pair are shown in Fig. 2.11 : a finger structure (Fig. 2.11(a)), an interdigitated finger structure (Fig. 2.11(b)), a common-centroid structure (Fig. 2.11(c)) and an interdigitated waffle structure (Fig. 2.11(d)). The mismatch of these structures has been measured and compared in [Bas 96b]. It was found that the interdigitated waffle and the common centroid (quad) transistor layout structures show no systematic mismatch, and that the matching follows the model described in the previous section. Under induced die stress due to packaging, the finger style transistor pair and the interdigitated finger structure show a fluctuation on β matching up to 5 times higher than predicted by the model. This can be explained by piezoresistive effects due to residual stresses induced into the silicon chip by packaging. It has been shown that this effect can significantly degrade the matching performance of MOS transistors [Bast 96c]. Layout structures with a 2-axial symmetry, like the common centroid structure, can be used to cancel out stress, process and thermal gradients in every direction.

- **Same Orientation**

  Anisotropic process steps cause asymmetries in process parameters and the silicon substrate itself can be anisotropic. The mismatch caused by these effects can be avoided by placing matching devices with equal orientations and such that the current flow is strictly parallel and in the same direction (see Fig. 2.12).

- **Same Surroundings**

  Devices with un-identical surroundings can show a considerable mismatch. There are various possible reasons for this effect, which are not always clear. This problem can be solved by implementing dummy devices to simulate the same surroundings (see Fig. 2.13).

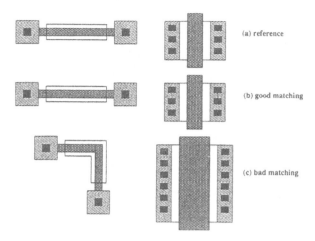

Figure 2.10: Influence of device shape on matching.

2.8   Thermal Effects

A current trend in IC technology is the integration of power devices together with complex
digital and high-performance analog circuits on the same chip. Examples can be found in such
diverse applications as smart power IC's, telecommunication circuits and high-speed memories
and microprocessors. The increasing power densities on these chips are reaching levels where
parasitic thermal effects limit performance. In this section we describe the influence of thermal
parasitics on the performance and reliability of analog circuits. We briefly discuss the influence
of temperature on device parameters and on circuit level performance characteristics. After that,
we discuss thermal models for integrated circuits.

2.8.1   Effect of Operating Temperature on Electrical Parameters

It is well known that temperature has an important effect on the electrical parameters of integrated
circuit devices [Laker-Sansen 94, Host 85]. Both the threshold voltage and the transconductance
of a MOS transistor depend on temperature [Laker-Sansen 94]. The primary variation of the
threshold voltage occurs through the temperature dependence of the Fermi potential ϕ_F. The
transconductance parameter KP depends on temperature because of the mobility, which can be
a strong function of temperature, depending on the doping level. A bipolar transistor consists
of two diodes, which makes its behavior strongly temperature dependent as well. The high-
precision capacitors and resistors which are often used in analog integrated circuits also show a
considerable temperature dependence [Cheng 89].

The temperature dependence of device parameters influences both the long-term reliability
and the performance of integrated circuits. Since the failure rate of micro-electronic devices dou-

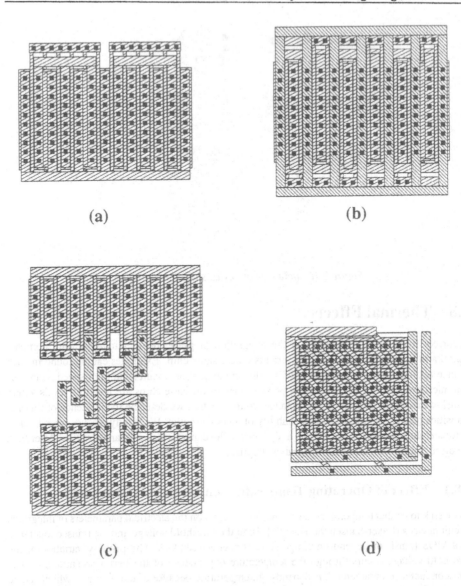

Figure 2.11: Matched transistor pairs.

bles for approximately every 10 deg C increase in temperature, hot spots due to excessive local power dissipation have become a major long-term reliability concern in many applications. In addition, the temperature dependence of device parameters results in thermally induced perfor-

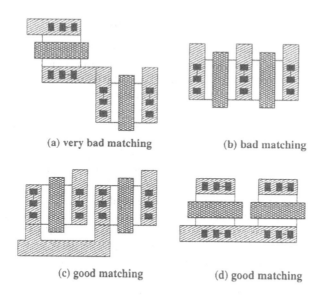

(a) very bad matching   (b) bad matching

(c) good matching   (d) good matching

Figure 2.12: Influence of device orientation on matching.

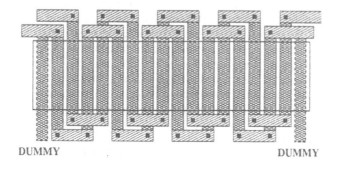

DUMMY   DUMMY

Figure 2.13: Matched resistor pair with dummy resistor strips added to improve matching.

mance degradation on the circuit level. This performance degradation can be due to a shift in absolute temperature of individual components, or can be caused by temperature gradients between components. An example of the first category is reference voltage shift in regulators and data converters [Fuka 76]. The second category includes input offset voltage, offset voltage drift and unwanted dc-feedback in differential amplifiers [Sol 74]. To control this thermally-induced performance degradation, it has become essential to take thermal effects into account during layout design.

2.8.2 Thermal Analysis of Electronic Systems

The source of heat in integrated circuits is current resistance. As heat is generated in a device, temperatures in the vicinity of the source will rise if the heat can not find a path from its source to a cooler sink. There are three ways for heat to flow away from a heat source to a heat sink : conduction, convection and radiation. Conduction heat transfer occurs through a solid when hot molecules transfer some of their vibrating, thermal energy to adjacent cooler molecules. The conduction heat transfer between any two points in an integrated circuit is proportional to the temperature difference between them, and inversely proportional to the thermal resistance of the silicon substrate. Convection heat transfer is heat exchanged between a solid and a fluid, such as air, at the solid-fluid boundary. The direction of heat transfer is from the solid to the fluid if the fluid is cooler than the solid and vice-versa if the fluid is hotter. Radiation heat transfer is the exchange of radiant heat energy between two separated bodies.

Conduction, convection and radiation mechanisms control the thermal behavior of an electronic system. From a systems perspective, an electronic assembly has to be modeled as a complex heat exchanger in which the power dissipated by the components is transported via these three heat transfer mechanisms to some ultimate heat sink. For chip level thermal analysis, a simplified model can be used.

Fig. 2.14 shows a schematic representation of a multi-layer electro-thermal integrated circuit structure with a central square surface heat source on top. The different layers model the different parts of the integrated circuit structure, such as the different silicon layers and the die-bonding layer. We assume that the mounting surface of the structure is isothermal and that its temperature has been determined through system level thermal analysis. In addition, we also assume that the open surfaces of the structure are adiabatic. This means that we neglect heat losses due to radiation and convection and therefore, assume heat flow to occur only by conduction. We also neglect heat loss through the wires bonded to the semiconductor chip. These assumptions are supported by experimental results reported in [Geer 93, Gray 71].

Within every layer, the material is assumed to be linear, isotropic and homogeneous. Temperature and heat flow are continuous at interfaces between layers. Power is generated uniformly in the surface source with power density Q_0''. The dimensions of the chip and the source are denoted by L_x, L_y and w, h respectively (see Fig. 2.14).

To obtain the static temperature distribution for this problem, the following heat-flow equation has to be solved :

$$\nabla^2 T(x, y, z) = 0 \tag{2.40}$$

with boundary conditions :

$$k_i \left. \frac{\delta T_i}{\delta x} \right|_{x=\pm \frac{L_x}{2}} = 0 \qquad i = 1, 2, \ldots, N \tag{2.41}$$

$$k_i \left. \frac{\delta T_i}{\delta y} \right|_{y=\pm \frac{L_y}{2}} = 0 \qquad i = 1, 2, \ldots, N \tag{2.42}$$

$$T_1 |_{z=-L_z} = 0 \tag{2.43}$$

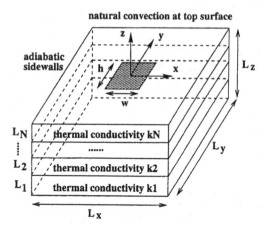

Figure 2.14: Multi-layer thermal model for an integrated circuit device

where $L_z = \sum_{i=1}^{N} L_i$ is the total height of the structure. Power is generated uniformly in the surface source with power density Q_0'' :

$$k_N \left. \frac{\delta T_N}{\delta z} \right|_{z=0} = p(x, y), \tag{2.44}$$

with:

$$p(x, y) = \begin{cases} Q_0'' & \text{for } |x| \leq \dfrac{w}{2}, |y| \leq \dfrac{h}{2}, \\ 0 & \text{elsewhere} \end{cases} \tag{2.45}$$

With these assumptions, the temperature distribution on the top surface of the N-layer structure can be written as a double Fourier series [Alb 95]:

$$T(x, y) = \frac{4Q_0''}{k_N} \sum_{n=0}^{\infty} \sum_{m=0}^{\infty} \tau_N(n, m) \cdot \frac{\sin \dfrac{n\pi w}{L_x}}{(1 + \delta_{n0})n\pi} \cdot \frac{\sin \dfrac{m\pi h}{L_y}}{(1 + \delta_{m0})m\pi} \cdot \cos \frac{n\pi x}{L_x} \cdot \cos \frac{m\pi y}{L_y} \tag{2.46}$$

The Fourier coefficients $\tau_N(n, m)$ depend upon the thermal conductivities k_i and thicknesses L_i of the layers in the multi-layer structure and can be computed with the following recursive relation :

$$\tau_1(n, m) = \tanh(\gamma(n, m)L_1) \tag{2.47}$$

$$\tau_N(n, m) = \frac{k_{N-1}\tanh(\gamma(n, m)L_N) + k_N \tau_{N-1}\gamma(n, m)}{k_{N-1} + k_N \tau_{N-1}\tanh(\gamma(n, m)L_N)} \tag{2.48}$$

$$\gamma(n, m) = \sqrt{\left(\frac{n\pi}{L_x}\right)^2 + \left(\frac{m\pi}{L_y}\right)^2} \tag{2.49}$$

Using (2.48) and (2.49) to compute the Fourier coefficients, it is straightforward to compute (2.46) for the two-layer and the four-layer case. These results are in exact formal agreement with the expressions derived in [Van Pet 93] for the two-layer case and in [Kokkas 74] for the four-layer case. The two-layer results derived in [Van Pet 93] have been experimentally verified using a thermal measurement procedure based on a BiCMOS diode matrix [Van Pet 93, Geer 93]. This and other studies [Lee 88] have shown that, when the heat source edges are at least one structure thickness away from the boundaries of the rectangular structure, the thermal profiles are weakly affected by the boundaries and thus the boundaries can be assumed to extend to infinity.

2.8.3   Discussion

The thermal solution (2.46) is expressed in terms of an infinite double Fourier series. In practice, however, only a finite number of terms can be taken into account. It was shown in [Lee 89] that the required number of terms in the series for a designated accuracy is directly proportional to the ratio of the chip to source size. For the analysis of structures with large chip-to-source size ratios, a considerable amount of CPU time is needed for (2.46) to converge. Consequently, it is impossible for thermally constrained layout tools to directly evaluate circuit level thermal performance degradation using (2.46). To overcome this problem, we have designed an efficient and reasonably accurate incremental thermal computation scheme based on (2.46). This computation scheme will be discussed in detail in chapter 4.

2.9   Substrate Coupling

In modern VLSI circuit design more and more analog and digital functions are realized on a single chip. In these mixed-signal chips, noise generating digital circuits and noise sensitive analog circuits are implemented on the same substrate. A result of this higher level of integration is the increased interaction between the circuits through the common substrate. Since the isolation provided by the substrate is non-ideal, current can flow through the substrate and couple circuits located in different parts of the chip [Johnson 84, Su 93, Smedes 93a, Stanisic 94, Ghar 95a, Ghar 95b]. As will be discussed in this section, substrate coupling is a layout related effect and careful layout design is needed to minimize unwanted substrate coupling.

2.9.1   Injection, Reception and Transmission of Substrate Current

Substrate coupling occurs when current is injected in the substrate by one device, transmitted through the substrate and picked up by another. A brief overview of these mechanisms is given in this section. Detailed descriptions can be found in [Su 93, Ghar 95b].

The current injection and reception mechanism depends on the type of device. Most active and passive devices inject current through their parasitic substrate capacitances. Examples are the source/drain to substrate capacitance of MOS transistors, the collector to substrate capacitance of vertical bipolar transistors, parasitic substrate capacitance of the bottom plate of a metal-oxide-metal capacitor. This capacitive coupling can be limited by putting devices in a grounded well.

If the well potential is allowed to vary with respect to ground, the well itself acts as an injector of current. Another significant current injection mechanism is impact ionization in MOS transistors.

The reception of substrate noise occurs mostly through capacitive sensing. The parasitic substrate capacitances that act as injectors of substrate current can also act as receiver. In MOS transistors, the body effect is also a severe form of substrate interaction. The body effect makes the drain current dependent on the substrate potential and causes a MOS transistor to pick up local substrate potential variations. While the capacitive coupling effects become significant only at relatively high frequencies, the body effect can be an issue at low frequencies.

For a first order calculation of substrate coupling effects, it is adequate to consider the substrate as a distributed resistance. This approximation is valid for frequencies up to $2GHz$ [Ghar 95b]. At frequencies above 4-5 GHz, the error in this assumption may become too large and a correct model of the substrate has to be obtained by solving Maxwell's equations in the substrate. For todays designs however, transmission of substrate noise is mainly due to the voltage variations induced by the injected current over the distributed substrate resistance.

2.9.2 Modeling of Substrate Coupling

To study substrate coupling effects, the substrate can be modeled as an n-port resistive network, where n is the number of contacts to the substrate. A contact is defined as a region where the circuit interacts with the substrate (see section 2.9.1). Computing a substrate model requires the computation of all resistance elements in the matrix model. This problem is similar to the capacitance and resistance extraction problems for interconnects and the same techniques that are used in interconnect modeling are also used in substrate modeling.

The first substrate modeling technique was introduced in [Johnson 84], where large 3-D resistor networks were used to study substrate coupled switching noise. In [Su 93], the large resistor network is replaced by a single node to model substrates which use an epitaxial layer grown on a heavily-doped bulk wafer. In [Verghese 93], a finite difference modeling technique was employed to generate substrate macro-models. 3D boundary-element methods are also used and offer the advantage of generating much fewer elements [Smedes 93a, Smedes 95, Ghar 95b].

Because of their computational requirements, the numerical methods that are mentioned above are limited to handle circuits containing a few hundreds of substrate contacts. In practice however, substrate effects occur in large mixed-signal circuits. Therefore, several attempt have been reported to speed-up the computation of substrate resistances. In [Joarder 94] parameterized lumped models are used to model several different isolation schemes using guard-rings. In [Verghese 95], precomputed point-to-point impedances and several other techniques are used to speed up the computation of the substrate admittance matrix. Another simplified substrate modeling technique was introduced in [Genderen 96]. The method involves the use of a common substrate node to which all contacts are connected via a resistance. Direct coupling resistances are only computed between neighboring terminals.

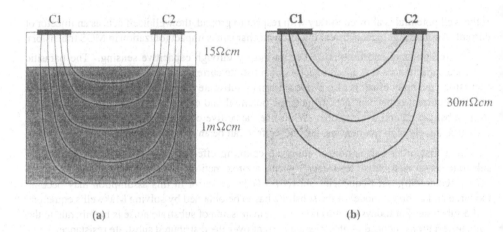

Figure 2.15: The flow of substrate current in different substrate types
(a) low impedance substrate
(b) high impedance substrate

2.9.3   Layout Measures to Reduce Substrate Coupling

2.9.3.1   Physical Separation

One of the approaches that can be taken to improve the isolation between noisy and noise-sensitive circuits is to physically separate the two. The effectiveness of this technique depends heavily on the type of substrate that is used to implement the circuit.

Two types of substrates are currently in use for industrial integrated circuit manufacturing : a low impedance and a high impedance substrate [Su 93]. The low impedance substrate consists of a lightly doped p or n-type epi layer on top of a heavily-doped p or n-type bulk. The high impedance substrate has a uniform lightly-doped bulk region. The type of substrate has a crucial impact on the substrate noise effects. In a low impedance substrate, most of the lateral current flow is through the heavily doped bulk (see Fig. 2.15(a)). This bulk region behaves electrically as a single node and therefore, the physical separation between noise injector and receiver has no influence on the magnitude of the coupling. Experimental results and device simulations indicate that when a switching noise source and a sensitive analog circuit are separated by more than four times the effective thickness of the epitaxial layer (i.e. $\pm 40 \mu m$) substrate crosstalk occurs through the heavily doped bulk and no improvement is obtained by increasing the separation [Su 93].

In high impedance substrates, the current flow is more uniform within the substrate because there is no low-resistance bulk (see Fig. 2.15(b)). Therefore, the isolation between digital and analog circuits improves as the physical separation is increased. This effect of the substrate type on the current flow was simulated and experimentally verified in [Su 93]. It was found that the magnitude of the coupling decreases almost linearly with the separation distance. Hence, for this

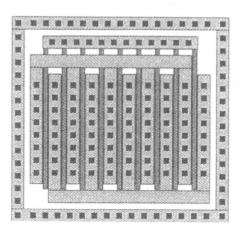

Figure 2.16: A MOS transistor protected by a guard ring

type of substrate the performance of the chip can be improved by separating noisy digital and noise-sensitive analog circuitry.

2.9.3.2 Guard Rings

Another commonly used layout measure is the use of guard rings around the injectors and the receivers of substrate currents in the circuit. Guard rings are substrate contacts that completely enclose a given region. They are connected to ground and provide isolation by absorbing the substrate potential fluctuations generated by the other devices. This technique has been shown to be effective only when the guard rings are placed very close to the sensitive analog circuits and are biased using dedicated package pins. In that case, they have the effect of creating a zero potential ring around the sensitive device, thereby electrically isolating it from the rest of the circuit.

2.10 Summary and Conclusions

The performance of analog integrated circuits is influenced by the parasitics that are introduced during layout. Three major categories of layout parasitics have been discussed in this chapter : interconnect parasitics, device mismatch and thermal effects.

Interconnect parasitics are the most important source of performance degradation in integrated circuits : they reduce the frequency bandwidth and introduce unwanted signal coupling. Since these parasitics are unavoidable, the task of a performance driven layout tool is to distribute them over the nodes of the circuit such that their combined impact on the performance is kept within the circuit specifications. Interconnect parasitics must be considered during each phase of the layout generation process. During module generation, the device level interconnect parasitics are fixed. The placement algorithms determines the minimum values of the interconnect parasitics while the router fixes their final values.

A second performance degrading layout effect is device mismatch. This mismatch can be limited by keeping the layouts of matched devices as identical as possible : equal orientations, shapes and equal surroundings. The distance between matched devices has to be traded of with other parasitic effects during placement in order to achieve the best overall performance. Special layout structures can be used if a very high matching degree is required.

Thermal effects are a third category of performance degradation in integrated circuits. Power dissipated by devices in a circuit causes thermal gradients across the chip. These gradients result in temperature shift for individual devices and thermal mismatch in matched device pairs. Thermal effects can be limited by careful placement of devices.

The goal of a performance driven layout system is to generate a layout such that the combined effect of these three major layout parasitics on the performance of a circuit remains within the specifications. In this chapter we have described a performance driven layout strategy that achieves this by driving the layout tools directly by the performance constraints. The cost functions used to quantify intermediate place and route solutions are based on an evaluation of the performance degradation that would result from accepting the solution : when it exceeds the allowed margins, the solution is penalized. In the remainder of this work, this methodology will be applied to module generation, placement and routing.

Chapter 3

Module Generation

3.1 Introduction

According to the macro-cell layout strategy presented in the previous chapter, the circuit schematic is first divided in modules, which are then optimally placed and routed. A module itself is defined as a functional set of one or more devices. For each module, a set of layout alternatives, called *variants* has to be generated. The placement tool then selects an optimal variant for each module, such that the overall placement is optimal in terms of area and performance. In this chapter we discuss the problem of circuit partitioning, i.e. the division of the circuit into modules, and module generation, i.e. the generation of a set of layout alternatives for each module.

Section 3.2 gives a description of the problem and some important issues which have to be taken into account during module generation. An overview and comparison of module generation strategies is given in section 3.3. Two recently proposed transistor stacking algorithms are discussed in section 3.4. The design and implementation of the LAYLA module generator library is the subject of section 3.5. In section 3.6 we describe the techniques that are used to make the library technology independent. We end this chapter with some examples in section 3.7 and we present conclusions in section 3.8.

3.2 Problem Formulation

The analog module generation problem can be formally stated as follows : given an electrical circuit specified as a set of devices (transistors, capacitors and resistors) and a netlist interconnecting these devices, divide the circuit into groups of devices (modules), such that each device is part of one and only one group, and a module generator is available to generate the layout for each module.

An important feature of analog integrated circuits is that each module can be laid out in different ways, even for the same parameter values: each module has a number of possible physical realizations, called variants, and each variant can have a number of possible aspect ratios or shapes (see Fig. 3.1). Depending on the actual parameter values of all modules and the

performance constraints, certain variants and shapes are more appropriate than others to obtain
a functional and still dense final layout within the user-supplied performance and geometrical
constraints. Therefore, the actual variant and shape of each module is only decided during the
placement (if not imposed by the user). It is the task of the module generators to generate the
palette of alternative variants for each module and the task of the placer to select the optimal one.

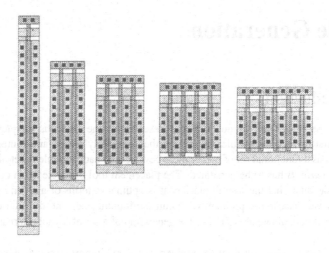

Figure 3.1: Different variants for the same MOS transistor.

In CMOS circuits, device capacitances can often be minimized by proper layout. The *pn*
junctions which form the MOS device source and drain regions each have a nonlinear voltage
dependent capacitance which consists of two terms, proportional to the junction area and perimeter, respectively :

$$C_{jSBt} = \frac{A_S C_j}{(1 - \frac{V_{BS}}{\phi_j})^{m_j}} + \frac{P_S C_{jsw}}{(1 - \frac{V_{BS}}{\phi_j})^{m_{jsw}}}. \tag{3.1}$$

In equation (3.1), C_{jSBt} refers to the total source bulk capacitance, A_S and P_S to the source area
and perimeter, C_j and C_{jsw} to the bottom and sidewall junction capacitances in absence of any
junction voltage and ϕ_j to the built-in junction potential. m_j and m_{jsw} depend on the doping
profile of the junction. Equation (3.1) reveals that this capacitance can be reduced by minimizing
the size of all diffusions. A large saving in diffusion capacitance can be made by *device merging*,
i.e. placing devices such that diffusion geometry is shared between electrically connected devices
as shown in Fig. 3.2. Two MOS transistors that have a common drain or source node (e.g. two
transistors in series have one common node, two transistors in parallel have two) can share a
diffusion region if they are of the same type and have the same bulk potential. This type of

geometry sharing not only improves the density of a layout, but also the performance of the circuit by reducing the parasitic capacitance of the node which has been merged.

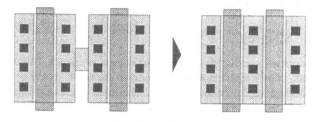

Figure 3.2: Device merging.

3.3 Module Generation Strategies

There are several ways of handling geometry sharing optimizations in analog circuit level layout generation. They can be incorporated in the module generation process or constructed dynamically during placement. A combination of both techniques can also be used.

3.3.1 Fixed Library of Procedural Generators

The first approach incorporates geometry sharing optimizations in a set of procedural module generators [Kayal 88, Rijm 89, Mogaki 89, Meyer 93]. In this approach, the circuit is first divided into a set of frequently encountered multi-device subcircuits, such as differential pairs, current mirrors, cascode stages, etc., and a dedicated procedural module generator is used to generate the layout for each type of subcircuit. A procedural module generator is a small parameterized program that generates a layout for a fixed configuration of devices. Since the topology of the devices within the subcircuit is fixed, geometry sharing optimizations can be hard-coded into the generator. As explained in the previous section, this improves both the layout density and the performance of the circuit. In addition, it reduces the complexity of the placement and routing problem, since devices are grouped in advance.

The disadvantages of this approach are related to the non-trivial problem of mapping the circuit onto an available set of module generators. Since the module generator library can only cover a limited set of sub-circuits, situations can arise where a potential geometry sharing situation allowed by a circuit topology does not match one of the pre-defined module generators. In this case, the opportunity for geometry sharing is lost and the performance and layout density suffer. Alternately, it is sometimes the case that a particular set of devices can be mapped into available module generators in more than one way.

Analog layout systems that employ a fixed library of module generators have to rely on the designer or on module recognition techniques to construct an adequate module structure. Hard-coded module recognition techniques are used in [Kayal 88, Mogaki 89, Rijm 88].

In [Meyer 93], the partitioning process can be controlled by a set of user defined rules. Another disadvantage of this approach is that the library of module generators has to be maintained whenever the technology process is updated. Procedural module generation is discussed in secion 3.5.

3.3.2   Dynamic Merging

In [Cohn 91] an alternative approach to geometry sharing was proposed. This approach relies on a rather simple library of procedural module generators to create the geometry for *single devices* (MOS and bipolar devices, capacitors and resistors). Merged structures are assembled from these smaller components by the placement algorithm. In contrast with the static approach discussed in the previous section, this approach explores the possible geometry sharing optimizations dynamically during placement.

The advantage of this technique is that it considers *all* possible geometry sharing optimizations during placement. The impact on the layout density and the performance of the different geometry sharing optimization possibilities can be incorporated into the cost function of the placement algorithm. This allows the placer to select the best possible geometry sharing configuration out of the different alternatives.

However, this technique also has a number of disadvantages. First, it places the burden on the placement algorithm. Detecting and constructing merged layout structures is now the responsibility of the placer, so a much more complicated program is needed. Incorporating geometry sharing optimizations into the placement algorithm increases the CPU time of the algorithm. Second, no matter how complicated the algorithm, some structures can never be constructed using geometry sharing techniques (e.g. interdigitated cascode structures). So, to obtain the best results, a number of subcircuit generators will have to be constructed anyway.

3.3.3   Simultaneous Placement and Module Optimization

In [Charbon 94a, Charbon 94b], simultaneous placement and module optimization is proposed as a way to explore geometry sharing optimizations. First, a stack-module generator [Malavasi 95] is used to partition a circuit and to find different alternative sets of modules. Each alternative set contains all the transistors of the circuit, split into modules merged with each other. All possible sets are generated by the algorithm and a cost function is used to select the best alternatives. The cost function takes into account constraints on parasitic junction capacitances, area, local routability, matching and symmetry constraints.

Next, a simulated annealing algorithm is used to generate a placement. One of the moves that is used to perturb the placement is a swap between alternative sets of modules. Such a module swap move replaces the entire subcircuit associated with the module with a new one, selected out of the set of alternatives which has been determined in advance during the module generation phase.

The technique can be seen as a compromise between efficiency and optimality. Instead of considering all possible geometry sharing configurations during placement, an optimized subset is determined in advance, and only this subset is considered during placement optimization. In this way, the efficiency and the robustness of the algorithm is improved.

3.3.4 Discussion

Based on experiments with a large number of industrial analog circuits, we have found that neither of these strategies does the job in an optimal way. A number of substructures returns over and over again in analog circuits and can not be assembled by merging elementary devices. Examples include interdigitated casdoded gain stages and common centroid differential input pairs. For these components, specific procedural module generators have to be written.

These module generators have to be combined with merging during placement to achieve close to optimal layouts in a fast, reliable and predictable way. In LAYLA, we have used a dynamic merging strategy, similar to the one presented in [Cohn 91]. In our placement algorithm, potentially beneficial overlaps between device source/drain regions are explored at every iteration. When such an overlap situation occurs, the total junction capacitance of the net to which the terminals belong decreases with an amount proportional to the area and the perimeter of the overlap region. This decreases the total performance degradation and lowers the cost function. In this way, device geometry sharing is promoted specifically for sensitive nets.

To overcome the problem of technology dependent module generator libraries, we propose a technique that isolates technology dependence in technology parameter files and device definitions (see section 3.6. This techique allows us to reuse the module generator library for different technologies without modifications to the source code.

3.4 Transistor Stacking Algorithms

In this section, we describe two algorithms that have been proposed for the automatic generation of an optimal set of transistor stacks implementing a CMOS analog circuit [Malavasi 95, Bas 96]. *Transistor stacking* or *transistor chaining* is defined as merging the diffusion regions of two or more transistors that have a common node. The layout structure that is formed by merging transistors is called a *stack* or *chain*. Transistor stacking algorithms have been studied intensively in the context of digital CMOS leaf cell generation [Uehara 81, Wimer 87, Chak 90]. The two analog stacking algorithms that were proposed in [Malavasi 95, Bas 96] use the result of the research in digital leaf cell generation, adding analog constraints like symmetry and matching, and using a special cost function to enforce parasitic constraints.

To solve the transistor stacking problem, a CMOS circuit is transformed into a circuit graph or diffusion graph $G(E, V)$, where each vertex is a net of the circuit. For each MOS transistor in the circuit, an edge is inserted in E, linking the vertices associated to the nets connecting to the source and drain terminals of the transistor. Fig. 3.3(a) shows the schematic of a CMOS OTA and the corresponding circuit graph is shown in Fig. 3.3(b). A full-stacked implementation of the layout corresponds to a path partition of the circuit graph, such that each edge in the graph belongs to one and only path. To generate a set of stacks for the circuit, the following steps are executed [Malavasi 95, Bas 96] :

1. The circuit graph is divided into a number of subgraphs, such that each subgraph contains only transistors of the same type, with the same bulk bias net (see Fig. 3.3(c)).

2. Large transistors are split into smaller parallel transistors (*device folding*).

3. Each subgraph is split into smaller connected subgraphs, containing only edges corresponding to transistors with the same channel width.

4. An optimum set of stacks is generated for each subgraph independently.

These four steps are common to the two algorithms that will be discussed. The difference between the two concerns step 4, stack generation for a subgraph.

In [Wimer 87] and later in [Malavasi 95] for analog circuits, stack generation is done in two phases. In the first phase, all the paths in the circuit graph are generated by a dynamic programming procedure. In the first step of this iterative procedure, all paths of length 1 are generated. In the n-th step, the paths of length n are generated by augmentation of the paths of length $n - 1$. In the second phase, a *path-graph* G_p is generated. The paths generated in the first step become the vertices of G_p, and an edge is inserted between two vertices if the corresponding paths are *compatible*, i.e. if they can coexist in the same partition. A circuit partition corresponds to a *clique* of G_p, i.e. a maximally connected subgraph of G_p. All circuit partitions are then generated using a clique finding graph algorithm. The quality of each clique (partition) is evaluated using an analog specific cost function. The cheapest clique found is the optimum solution to the partitioning problem. This algorithm has exponential time complexity [Malavasi 95]. The reason for this is that it enumerates all stacks, which makes it extremely sensitive to the size of the problem.

In [Bas 96] a constraint-driven stacking algorithm with linear time complexity was proposed. It employs an Eulerian trail finding algorithm that can satisfy analog-specific performance constraints. A *trail* on a graph is a set of edges $(v_0, e_0, v_1, e_1, \ldots, v_{k-1}, e_{k-1}, v_k)$, where $e_i = (v_i, v_{i+1})$ is an edge in the graph and $e_i \neq e_j$ for all $i \neq j$ [Thul 92]. The trail is a *closed trail* if $v_0 = v_k$. A closed trail is an *Eulerian* trail, if it touches all the edges in the graph. If a closed Eulerian trail exists in a graph, the graph is called Eulerian. A graph is Eulerian if and only if it is connected and all vertices in the graph have even degree [Thul 92]. An Euler trail in a circuit graph defines the optimal stack of the circuit. The algorithm proposed in [Chak 90] and modified for analog constraints in [Bas 96], proceeds in two steps. First, if the circuit graph is not Eulerian, it is made Eulerian by adding an extra vertex (*supervertex*) and an edge between the supervertex and every odd-degree vertex. The resulting graph is called modified circuit graph. Every edge that is added to the original circuit graph will result in a gap in the diffusion area of the stack. In the second step, an Eulerian trail finding algorithm is run on the modified circuit graph. The Eulerian trail finding algorithm in [Bas 96] is adapted to support symmetry and matching constraints and works with a constraint driven cost-function. The complexity of the algorithm is linear in the number of transistors in the circuit and guarantees to find the minimum cost stack set for a certain class of circuits [Bas 96].

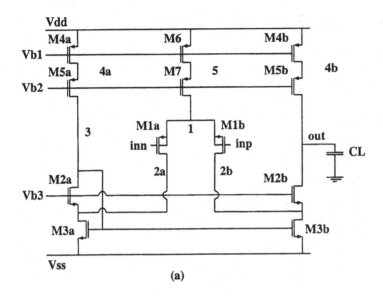

(a)

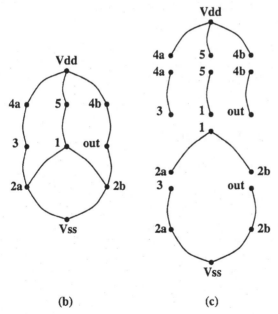

(b) (c)

Figure 3.3: Highspeed OTA :
(a) circuit schematic,
(a) circuit graph,
(b) circuit subgraphs, with equal bias net

3.5   Procedural Module Generation

A parameterized module generator is a program that procedurally generates a layout for a fixed module topology, based on a set of module parameters, a technology specification and a number of user specified options. In general, layouts created by module generators can be of any complexity, ranging from basic devices (transistors, capacitors, resistors) to entire subsystems (e.g. RAM blocks). In the context of analog circuit level layout, their complexity is usually restricted to one or a few devices.

Module generators can be used as stand alone tools during interactive layout sessions or in the context of an automatic placement program. In the latter case, they have to support two different operating modes :

1. **interface mode** : when used in interface mode, the module generators take as input the actual parameters for the module and return a collection of geometrical abstractions for all possible variants that can be used to lay out the module. The geometrical abstraction can be as simple as a bounding box or can be more complicated, depending on the information that is needed by the placement algorithm to evaluate intermediate solutions. The details of geometrical abstraction that is used in our placement model will be discussed in chapter 4.

2. **layout mode** : when working in layout mode, the module generator constructs a detailed physical layout for the actual variant selected by the placement algorithm.

In the beginning of the placement process, the generators are used in analysis mode to obtain information about the different possible implementations of a module. When the placer is finished, the actual variant is known, and the generators can be used in synthesis mode to obtain the actual physical placement.

3.6   Technology Independence

Writing module generators is a tedious and labour-intensive task and generator libraries turn out to be large software systems. It is therefore crucial that module generators are written in a process-independent way, to make it easy to port them to new technologies.

Technology independence is achieved by writing the module generation procedures in terms of symbolic process parameters and layers. The values of these parameters and layers for a particular technology process are stored in a separate technology file. When the module generator is called for a certain technology process, the appropriate technology file is read in and all technology parameters are instantiated. In this way, technology specific information is kept out of the actual module generation code and the generators can be ported to new processes by writing new technology files.

A module generator is basically a program that generates a collection of polygons on layers. The dimensions of the polygons depend on the module parameters and on the design rules. The layers on which they are instantiated depend on the process mask sequence. To achieve real technology independence, both types of process specific information (design rules and process layers) have to be parameterized in the module generation code. We have introduced *process parameters* and *module definitions* to achieve this.

- Process Parameters

  The process parameters can be classified in two categories :

  - **Geometrical Process Parameters** For every relevant design rule, a process parameter is defined. The module generator code contains statements that specify polygon dimensions as a function of module parameters and process parameters. During initialization of the module generator, the values for the process parameters are retrieved from the technology file and together with the module parameters, they are used to calculate the polygon dimensions. Examples of geometrical process parameters are : the minimum width of a layer, minimum separation between layers, the minimum enclosure of one layer with respect to another layer, etc.

  - **Electrical Process Parameters** Electrical Process Parameters define the electrical characteristics of the process layers. Typical examples are the sheet resistance and parallel plate capacitance of a layer, the maximum current density of a layer, etc.

- Device Definitions

  The *device definitions* are introduced to describe in a general way the mask layout structure of different devices in a particular process. For each device used in a module generator, a generic layer model has been constructed and the module generators have been written in terms of these generic layers. The device definitions, which are part of the technology file, are used to establish the correspondence between these generic layers and the layers used in a particular process. A 'void' layer is used if there is no corresponding layer in the process. The module generators can be adapted to different technology processes by writing new device definitions. An additional advantage of using device definitions is that it allows a user to define different types of transistors, capacitors and resistors. Device definitions are currently supported for contact structures, MOS transistors, capacitors and resistors. They are discussed in detail in [LAYLA man](Ch. 9).

3.7 Examples

The LAYLA module generator library currently consists of generators for MOS transistors, capacitors, resistors, inductors, transformers, pairs of MOS transistors, pairs of capacitors and pairs of resistors. The library is currently in use in the ESAT/MICAS group and in several electronic companies. Technology files have been written for more than 20 different processes from 5 different foundries. It takes about an hour to port the library to a new technology. Detailed information on the library can be found in [LAYLA man]. In the remainder of this chapter, we will present some examples to illustrate some features of the module generator library.

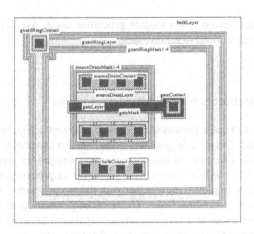

Figure 3.4: A generic MOS layout structure

3.7.1 MOS Transistor

3.7.1.1 MOS Definition

A MOSFET transistor is created each time a strip of polysilicon crosses an underlying diffused area. This description is, however, not sufficient to define a MOS layout structure in modern submicron processes. Additional layers are needed to define the well and the guard ring, or to enhance the transistors (e.g. to decrease the V_T of a PMOS transistor). A general structure of a MOS transistor is illustrated in Fig. 3.4. The structure consists of a gate layer, a source/drain layer and a guard ring layer. An optional mask layer can be defined for the gate layer and up to four optional mask layers for the source/drain and guard ring layers. In addition, the contact structures that have to be used to contact the gate, the source/drain and the guard ring have to be defined. Example 3.1 shows the device definition used to map the layers of the generic MOS structure to the layers used in a CMOS $0.7\mu m$ process. The comments give the names of the corresponding generic layers.

3.7.1.2 MOS Transistor Parameters

The following parameters can be specified as input for the MOS transistor generator :

- **width** : the channel width of the transistor.

- **length** : the channel length of the transistor.

- **fingers** : the number of *fingers* of the MOS transistor. Large transistors can be split into a number of parallel parts, which are then merged together. The parts are called fingers, and

```
mos_def(low_vt_pmos,      // mosType
        poly,             // gateLayer
        void,             // gateMask (not used)
        mlpoly,           // gateContactType
        active_area,      // sourceDrainLayer
        p_active_area,    // sourceDrainMask1
        p_diffusion,      // sourceDrainMask2
        low_vtp,          // sourceDrainMask3
        void,             // sourceDrainMaks4 (not used)
        mlpdiff,          // sourceDrainContactType
        active_area,      // guardRingLayer
        n_active_area,    // guardRingMask1
        void,             // guardRingMask2 (not used)
        void,             // guardRingMask3 (not used)
        void,             // guardRingMask4 (not used)
        mlndiff,          // guardRingContactType
        nwell,            // bulkLayer
        mlnwell,          // bulkContactType
        metal1,           // firstRoutingLayer
        metal2,           // secondRoutingLayer
        via);             // rountingViaType
```

Example 3.1: MOS transistor definition.

the process of splitting a transistor into parts is called *transistor folding*. Fig. 3.5 shows a folded transistor with 12 fingers.

- **slices** : the number of *slices* of the transistor. An effective way to protect sensitive devices against coupling noise is to make sure that no part of the transistor is more than a specified minimum distance away from a bulk contact. This minimum distance can be a design rule imposed by the foundry or can be specified by the designer. For extremely large transistors, this rule results in very long transistor layout structures. To avoid these inconvenient aspect ratios, very large transistors can be split into a number of parallel transistors, called *slices* and rows of bulk contacts can be inserted between them. Fig. 3.6(a) shows a large transistor which has been split into two slices. A row of bulk contacts has been added in between them.

- **current** : the current flowing through the transistor. If this current exceeds a process-specific threshold value, the widths of the source/drain wires have to be increased to avoid excessive voltage drops and electromigration effects. Fig. 3.6(b) shows a transistor layout with increased source/drain wire width. The number of vias used to connect metal 1 and metal 2 wires has also been adapted to the high current.

- **aspect ratio** : the aspect ratio of the layout of a transistor with a given W/L ratio is determined by the number of fingers and slices. Instead of specifying the number of fingers and slices, it is also possible to specify a desired aspect ratio. In this case, the number of fingers and slices will be determined, taking into account the bulk-contact and high-current layout rules.

- **type** : the MOS transistor type. The specified type must correspond to one of the device definitions that are defined in the technology file.

- **technology file** : the technology file that contains the design rules and device definitions to be used during generation of the transistor.

- **device name** : the name of the device.

- **generator mode** : the mode in which the transistor has to be generated (see section 3.5). This can be *interface* or *layout*.

- **output format** : different output formats can be specified, depending on the layout environment in which the generator is used.

3.7.1.3 MOS Transistor Examples

Fig. 3.5 shows a folded nMOS transistor with 14 fingers. An example of a large nMOS transistor that has been split into two parallel transistors is given in Fig. 3.6(a). Three parallel transistors are placed on top of each other and bulk contacts are inserted in between them.

The transistor in Fig. 3.6(b) has a high drain current. Therefore, the width of the horizontal and vertical source and drain wires has been increased as well as the number of vias used to connect the vertical source/drain wires to the horizontal ones. The width and the number of vias is calculated based on the maximum current per unit wire width and per via, which are specified in the technology file.

The width of each wire in Fig. 3.6(b) depends on the current flowing. Let I be the total drain current of the wire, then the width W_2 of the metal 2 wires connecting the slices together is given by :

$$W_2 = \frac{I}{I_{max,2}}, \tag{3.2}$$

where $I_{max,2}$ is the maximum current density of the metal 2 layer, expressed in Ampere per meter. If there are n_s slices in the transistor structure, then the width W_2 of the metal 1 wires connecting the fingers together is given by :

$$W_1 = \frac{I}{n_s I_{max,1}}, \tag{3.3}$$

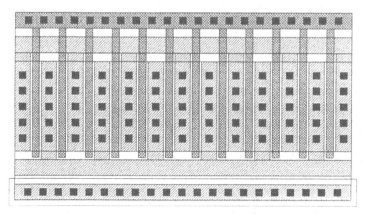

Figure 3.5: Automatically generated folded nMOS transistor layout.

with $I_{max,1}$ the maximum current density of the metal 1 layer. The neccesary number of vias N_{via} can be computed as follows :

$$N_{via} = \frac{I}{n_s I_{max,via_{12}}},$$ (3.4)

where $I_{max,via_{12}}$ is the maximum current in a via connecting layers metal 1 and metal 2. If there are n_f fingers per transistor slice, then the width W_{f1} of the metal 1 wires connecting the source/drain diffusion region is given by

$$W_{f1} = \frac{I}{n_s floor(\frac{n_f+1}{2}) I_{max,1}},$$ (3.5)

where $floor(x)$ denotes the largest integer smaller than or equal to x.

As a result of the wide source wires, the gate wires can become very long. This causes the gate resistance to increase which is bad for the noise performance and can lead to RC effects at high operating frequencies. The structure in Fig. 3.7 is used to solve this problem. In this structure, metal 2 is used for the source wires, which allows to put the gate wires closer to the active area of the transistor.

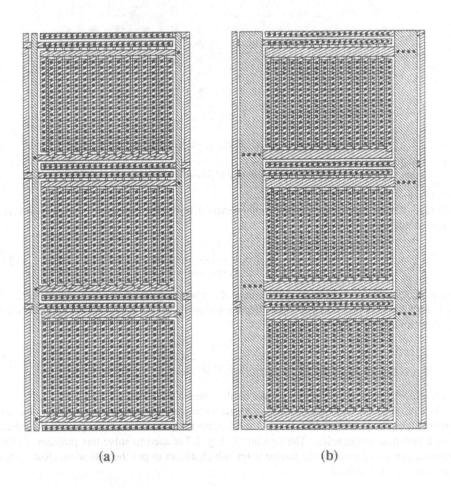

(a)   (b)

Figure 3.6: Layout structures for large MOS transistors :
         (a) minimum source/drain wire width
         (b) increased source/drain wire width for high current.

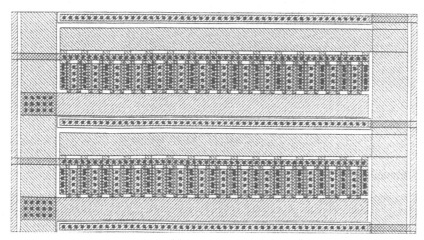

Figure 3.7: High current nMOS transistor, source wires in metal 2 to reduce RC effects

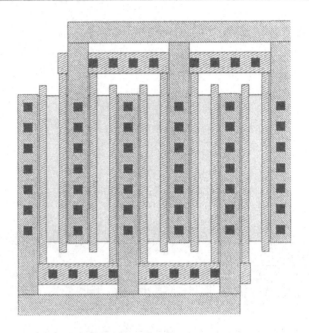

Figure 3.8: Cascode MOS transistor pair.

3.7.2   Cascode MOS Transistor Pair

Fig. 3.8 shows an example of a MOS transistor pair structure for which a dedicated module generator was written. This structure consists of two MOS transistors connected source to drain in a cascode configuration. No contacts are necessary on the diffusion regions that form the shared source/drain of the devices. This allows to put the poly gate wires at minimum distance which results in substantially reduced parasitic source/drain capacitance on this node. Note that this interdigitated structure can not be formed by the merging during placement technique described in section 3.3.2.

3.8   Summary and Conclusions

In this chapter, the circuit partitioning and module generation problem for analog circuits has been discussed.

   The layout of a CMOS circuit can be optimized with respect to area and performance by sharing source/drain diffusion regions among transistors that have a common node. This process is called geometry sharing optimization and the merged structures are called transistor stacks. Finding an optimal set of stacks implementing an analog circuit is not a trivial task, due to the large number of mutually exclusive geometry sharing possibilities which are usually present in analog circuits.

In general, three different strategies can be applied to module generation for analog circuit layout. The first strategy uses a large library of procedural module generators that implement most of the analog layout knowledge. The second strategy limits the number of procedural module generators as much as possible and relies on the placement tool to assemble the more complicated substructures. The third approach is to construct a limited number of promising alternative sets of stacks in advance and to select the best one during placement. Graph-based algorithms can be used to generate the palette of alternative stacked realizations.

It is our opinion that neither of these strategies does the job in an optimal way. A number of substructures returns over and over again in analog circuits and can not be assembled by merging elementary devices. For these components, specific procedural module generators have to be written. These module generators have to be combined with merging during placement to achieve close to optimal layouts in a fast, reliable and predictable way.

To overcome the problem of technology dependent module generator libraries, we have proposed a technique that isolates technology dependence in technology parameter files and device definitions. The LAYLA module generator library has been used with more than 20 different technologies from 5 different foundries without modifications to the source code. An example of a procedural module generator was presented to illustrate some analog specific features.

Chapter 4

Placement

4.1 Introduction

This chapter addresses the placement problem for high-performance analog circuits. The placement phase is crucial for the performance degradation of an analog circuit layout since it influences all the parasitic layout effects which have been discussed in chapter 2. The distance between matching devices, and therefore also their matching degree is determined during placement. The placement of a circuit also determines its thermal profile. In addition, it greatly influences the values of the interconnect parasitics. Although their final values are determined during routing, their minimum values are fixed by the configuration of the device terminals, which is determined during placement. A performance driven placement algorithm therefore has to take into account all of these performance degrading effects simultaneously.

We begin this chapter by formally stating the problem and reviewing the constraints that have to be taken into account during placement. We then give a brief overview of our placement tool in section 4.3. In order to justify our choice of simulated annealing as basic placement optimization algorithm, we give an overview and comparison of different placement techniques in section 4.4. Based on this comparison, the simulated annealing algorithm is selected and its application to analog performance driven placement is discussed in section 4.5. Some important details of our placement implementation, the placement model, the handling of analog constraints, the move set and cost function are discussed in sections 4.5.1, 4.7 and 4.8. In section 4.9 we describe in detail how the layout induced performance degradation is computed for an intermediate placement solution. In section 4.11 we discuss the annealing schedule of the algorithm. Finally, we give some experimental results in section 4.12 and we draw conclusions in section 4.13.

4.2 Problem Formulation

The analog circuit level placement problem can be stated as follows : given an electrical circuit specified as a set of devices (transistors, capacitors and resistors) and a netlist interconnecting terminals on these devices and on the periphery of the circuit itself, select an optimal implementation (variant) for each device, and position these variants on the layout surface in a design rule correct way and such that the layout area is minimal and that the circuit can be routed afterwards. The following additional constraints and objectives have to be added to this basic definition :

- **Symmetry Constraints**

  In high-performance analog circuits, it is often required that groups of devices are placed symmetrically with respect to one or more symmetry axes. Symmetric placement allows for symmetric routing and results in matched parasitics. Symmetry constraints can be formulated in terms of *couples*, *self-symmetric devices* and *symmetry groups*. Two devices which are placed symmetrically with respect to an axis form a couple. A self-symmetric device is a device which is placed on a symmetry axis. A symmetry group is a collection of couples and self-symmetric devices which share the same symmetry axis. The symmetry group represented in Fig. 4.1 consists of the couples (M1A,M1B) and (M2A,M2B) and the self-symmetric device M5. More than one symmetry group can be specified for a circuit. The presence of one or more symmetry groups has the following implications for analog placement :

  - Two devices which are specified as a couple must be placed symmetrically with respect to an axis and must have identical variants and mirrored orientations.

  - A device which is specified as self-symmetric must be placed on a symmetry axis.

  - Couples and self-symmetric devices that belong to the same symmetry group must share the same symmetry axis.

- **Matching Constraints**

  Matching constraints can be specified by defining matching groups. A matching group is a set of two or more devices for which an accurate ratio of device characteristics is required. The simplest and most common case of a matching group is a pair of equal devices. A more complicated case of a matching group is shown in Fig. 4.2. Any number of matching groups can be defined in an analog circuit. The presence of one or more matching groups has the following implications for analog placement :

  - All devices which belong to the same matching group must have equal orientations.

  - If the devices of a matching group are equal (1:1 ratios), they must be implemented with equal variants. If they have another ratio, they should be built of equal unit devices, according to the ratio.

  - The placement tool has to determine the positions and therefore also the distance between the matched devices such that the circuit performance constraints are met. Since it is not always possible in an analog circuit layout to, at the same time, meet all symmetry requirements, put all matching devices directly next to each other and obtain a fairly compact layout, the matching degree of a pair of devices has to be selected in view of its influence on the performance of the circuit.

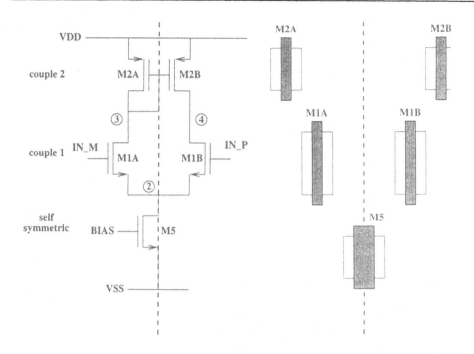

Figure 4.1: A circuit fragment with corresponding layout, illustrating symmetry constraints.

- **Performance Constraints**

  As discussed in chapter 2, the performance of a circuit is influenced by layout parasitics.

  - **interconnect parasitics**

    A performance driven placement algorithm has to create a placement that allows a router to complete the interconnections within the performance constraints. Although the actual values of interconnect capacitances and resistances are determined during routing, their minimum achievable values are fixed during placement and it is therefore crucial that (estimated) performance degradation induced by interconnect parasitics is taken into account during placement.

  - **device mismatch**

    The distance between the matching devices has to be selected in view of its influence on the performance of the circuit.

  - **thermal effects**

    The presence of power dissipating devices in a circuit causes a temperature distribution across the placement. Since device characteristics are influenced by local temperatures, matching devices have to be placed such that the performance degradation

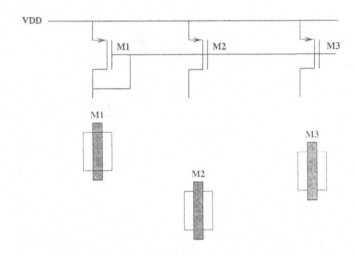

Figure 4.2: A circuit fragment with corresponding layout, illustrating a matching constraint.

introduced by their temperature difference remains within the specifications. Again, it is not always possible to place all matching devices exactly on isotherms, while at the same time meeting all other constraints.

- **Geometrical Constraints**

  The blocks generated by a circuit level layout generation tool are often part of a larger system. To minimize system level performance degradation, a target aspect ratio, fixed height and/or fixed terminal positions may be specified by a floorplanning tool or by a designer. These additional geometrical constraints also have to be taken into account during placement.

4.3 Overview of the Placement Tool

The basic components of the developed placement program are illustrated in Fig. 4.3. The input to the tool is the circuit netlist after sizing as well as the list of performance specifications that the circuit has to meet (e.g. phase margin $\geq 60°$). The difference between these specified values and the actual performance values obtained by the circuit after sizing (including statistical process parameter variations) are the margins that can be taken up by layout-induced performance degradation (see chapter 2). The technology file is also input to the tool.

The first step in the execution of the placement program consists of a number of numerical simulations which result in the set of required performance sensitivities and operating-point information (branch currents, node voltages) of the circuit. This information, together with the

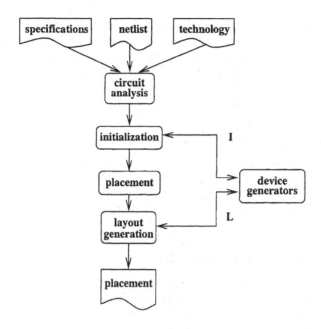

Figure 4.3: Overview of the analog placement tool.

circuit netlist, is then used as input for a set of module generators that construct a list of geometrical variants for every device (see chapter 3). Only the information needed for the optimization of the placement is generated (the generators are called in interface mode).

Next, a simulated-annealing algorithm is used to generate the actual placement, taking into account all the constraints and objectives that have been identified in the previous section.

After the optimization, the module generators are called again (this time in layout mode) to create the actual layout for the selected geometrical device variants and the final layout is constructed. The output of the program is then the final placement layout, together with information about the performance degradation in this final layout and an identification of the most important contributions to this degradation. In case the degradation exceeds the required performance specifications, this information can be used by the designer to see the failing performance(s) and to identify the critical effects. This allows him to improve his design when desired.

4.4 Previous Work in Placement Algorithms

Although the constraints identified in the previous sections make analog performance driven placement a unique problem, the core problem, arranging a set of connected blocks on a layout surface, has been studied extensively in the context of other related VLSI placement and floorplanning problems. These placement problems are known to be NP-complete [Sahn 80].

Because of this complexity, heuristic algorithms must be used to solve them. It would lead us too far to discuss all placement algorithms that have been used in the past. We refer the reader to the excellent overviews in [Shah 91, Sher 95] and the references therein for details on the various algorithms. In this section we will give an overview of the main classes of algorithms, with a short description of their working principle and main features. Based on this overview we will justify our choice of simulated annealing as basic algorithm for performance driven analog placement.

4.4.1   Constructive Placement (CP)

The earliest constructive placement techniques were based on connectivity information [Han 72]. They construct a placement by selecting a seed module and placing it on the layout area. The other modules are then selected one at a time and placed in the best available location. The order in which the modules are selected is based on connectivity information (most densely connected first), or on expert rules.

4.4.2   Force-Directed Placement (FDP)

Force-directed placement algorithms are rich in variety and differ greatly in implementation details [Han 72]. In general, they cast the placement problem into the classical mechanics problem of a system of bodies attached to springs. Blocks connected to each other by nets exert attractive forces on each other. The magnitude of these forces is directly proportional to the distance between the blocks. The ideal configuration of the placement of blocks is the one in which the system achieves equilibrium. Using this analogy, the placement problem becomes a problem of classical mechanics and the variety of methods used in classical mechanics can be applied to it. Some of these methods are constructive, others are based on iterative improvement.

4.4.3   Placement by Partitioning (PbP)

Partitioning based placement algorithms generate a placement by repeatedly partitioning a circuit into two sub-circuits such that the number of nets cut by the partition is minimized. At the same time, the available layout area is partitioned alternately in the horizontal and vertical direction and each sub-circuit is assigned to one partition of the layout area. This process is repeated until each sub-circuit consists of only one module and has a unique place on the layout area. Most placement by partitioning algorithms, or *min-cut algorithms* use some modified form of the Kernighan-Lin [Ker 70] or Fiduccia-Mattheyses [Fidu 82] heuristics for partitioning.

4.4.4   Quadratic Optimization (QO)

Quadratic optimization techniques   generate a minimum net-length placement by solving a quadratic minimization problem. Modules are represented by points which have to be placed on the layout surface. Complete-graph models are used to model a net. If the Euclidean (quadratic) distance norm is used to model the net-length, the problem of minimizing the total net-length can be solved as a quadratic minimization problem.

4.4.5 Simulated Evolution (SE)

Simulated Evolution emulates the natural process of evolution as a means of progressing towards an optimal placement solution. The algorithm starts with a randomly generated initial set of placement configurations, which is called the population. This population is iteratively improved using a procedure that mimics the natural process of evolution. First, the individuals in the current population are evaluated using some measure of fitness. Based on this fitness value, two individuals are selected as parents. The higher the fitness value of an individual, the higher the probability of being selected as a parent. A number of genetic operators (crossover, mutation and inversion) is then applied to the parents to generate new individuals, called offspring. The offspring is then evaluated and a new generation is formed by selecting some of the parents and offspring, based on their fitness. The fittest member in the final population represents the best placement solution found.

4.4.6 Simulated Annealing (SA)

Simulated annealing is a general optimization technique for solving combinatorial optimization problems. A combinatorial optimization problem is formalized in [Laar 87] as a pair (R, C), where R is the finite - or possibly countably infinite - *set of configurations* (also called *configuration space*) and C a *cost function*, $C : R \rightarrow \mathbb{R}$, which assigns a real number to each configuration. It is assumed, without loss of generality, that C is defined such that the lower the value of C, the better (with respect to the optimization criteria) the corresponding configuration. Solving a combinatorial optimization problem comes down to finding a configuration for which C takes its minimum value, i.e. a configuration i_0 satisfying

$$C_{opt} = C(i_0) = min_{i \in R} C(i), \tag{4.1}$$

where C_{opt} denotes the optimum (minimum) cost.

In condensed matter physics, annealing denotes a physical process in which a solid is heated up to a temperature at which all particles of the solid randomly arrange themselves in the liquid phase, followed by cooling through slowly lowering the temperature. If the maximum temperature is sufficiently high and the cooling is carried out slowly, the particles arrange themselves in the low energy ground state of an ordered crystalline lattice. The cooling has to be carried out sufficiently slowly to allow the solid to reach thermal equilibrium at each temperature value T. A solid is in a state of thermal equilibrium if its energy probability distribution is given by the Boltzmann distribution :

$$P\{energy = E\} = \frac{1}{Z(T)} exp(-\frac{E}{k_B T}) \tag{4.2}$$

where $Z(T)$ is a normalization factor depending on the temperature T and k_B is the Boltzmann constant. It can be seen from (4.2) that the probability of the low energy states increases as the temperature is decreased.

In [Met 53] the Metropolis algorithm was proposed to simulate the evolution to thermal equilibrium of a solid for a fixed value of the temperature T. In this method, the state of the solid, characterized by the positions of its particles, is repeatedly changed by applying a small random displacement to a randomly chosen particle. If the difference in energy between the old and the new state ΔE is negative, the new state is accepted and the process is continued with the new state. If ΔE is positive or equal to zero, the probability of acceptance of the new state is given by $exp(-\frac{\Delta E}{k_B T})$. This acceptance rule is referred to as the Metropolis criterion. Following this criterion, after a large number of perturbations, the system evolves to a state of thermal equilibrium, characterized by energy distribution (4.2). The Metropolis algorithm can thus be used to simulate the evolution of a solid to thermal equilibrium. By applying this algorithm at decreasing values of temperature, the annealing process of a solid can be simulated.

The simulated annealing algorithm is based on the analogy between the simulation of the annealing of solids and the problem of solving large combinatorial optimization problems. In the latter case, the configurations of the optimization problem play the role of the states of the solids and the cost function C associated with a particular configuration takes the role of the energy E of a state. A control parameter T is introduced to play the role of the temperature. The algorithm can now be described as follows. Initially, the control parameter T is given a high value and a sequence of configurations is generated using the Metropolis algorithm. Starting from the current configuration i, a new configuration j is chosen using a generation mechanism, i.e. a prescription to generate a transition from a configuration to another one by a small perturbation. If $\Delta C_{ij} = C(i) - C(j)$ is the difference in cost between the two configurations, the new configuration is accepted with probability 1 if $\Delta C_{ij} \leq 0$ and with probability $exp(-\frac{\Delta C_{ij}}{T})$ if $\Delta C_{ij} > 0$. This process is continued until equilibrium is reached, i.e. until the probability distribution of the configurations approaches the Boltzmann distribution, now given by

$$P\{configuration = i\} = \frac{1}{Q(T)} exp(-\frac{C(i)}{T}) \qquad (4.3)$$

where $Q(T)$ is a normalization constant depending on the control parameter T.

The control parameter is then lowered in steps, with the system being allowed to approach equilibrium for each step by generating a sequence of configurations in the described way. The algorithm is terminated for some small value of T, for which almost no deteriorations are accepted anymore. The final *frozen* configuration is taken as the solution to the optimization problem.

4.4.7 Discussion

Based on the problem description given in section 4.2, the following desirable features for an analog placement algorithm can be identified :

1. As pointed out in chapter 3, most of the devices in an analog integrated circuit can be laid out in different ways. Therefore, the algorithm should be able to select position, orientation and implementation (variant) simultaneously.

2. Analog devices can have arbitrary rectilinear shapes. Each device terminal can also have an arbitrary rectilinear shape. Device terminals in analog layout can not be reduced to points.

3. Most analog circuits are of moderate size. The average complexity of an analog circuit level placement problem is 20 to 30 devices. Although important, the efficiency of the algorithm is not as crucial as it is for large digital placement problems.

4. The sizes of devices encountered in analog circuits vary with orders of magnitude. Therefore, the algorithm should perform well with blocks of widely varying sizes.

5. The various symmetry and matching constraints that are frequently encountered in analog circuits, require accurate control over the devices positions and orientations. In addition to this, a number of geometrical constraints are often imposed on the overall layout (e.g. fixed height, aspect ratio, terminal positions, ...) Hence, a flexible placement optimization technique, that allows to arbitrarily constrain every aspect of the placement problem, is needed.

6. The most important objective in performance driven placement is to guarantee that the layout induced performance degradation remains within the circuit specifications. Therefore, placement decisions must be based on an accurate evaluation of the performance degradation, which requires detailed knowledge of the positions and orientation of all devices simultaneously at all times. Constructive and partitioning based algorithms take placement decisions sequentially, based on incomplete information and are therefore not well suited for analog placement problems.

If we take into account these criteria for the selection of an analog placement algorithm, the stochastic optimization algorithms (SE and SA) appear to be the most promising candidates. Both of these algorithms offer the possibility of putting arbitrary constraints on the generation of new candidate placement solutions. In SA this can be done through careful design of the move set, in SE through the use of restricted crossover, mutation and inversion operators. Both algorithms are cost function based. The cost function, that evaluates the quality (or fitness) of intermediate solutions can be used to implement the analog performance and geometrical constraints. SA and SE operate on the entire solution simultaneously, which is essential for an accurate parasitic evaluation. All these features combine to make SA and SE the best choice for analog placement problems.

One of the features that can be used to distinguish SA and SE algorithms is their interactivity control. SE operates on a pool of solutions simultaneously. This offers the possibility to the user to insert candidate solutions (*hints*) in this pool. Even if the hint solution is not accepted as the optimal one, it will continue to exist in the pool for some generations and the best aspects of the solution will influence the final solution through the crossover mechanism. This makes SE an ideal algorithm for placement problems where a great deal of user interaction is required (e.g. floorplanning algorithms).

Our analog placement tool was designed for 'batch' operation in an automated analog synthesis environment [Gielen 95a]. The main advantage of the SE algorithm is therefore not essential for our problem. In addition to this, SA for layout problems is much more mature than SE. A lot of research that has been done on the operation of the SA algorithm can be reused for our analog placement problem. This allows us to concentrate on the analog specific aspects of the problem, such as the implementation of the various constraints, and the evaluation of layout induced performance degradation. Therefore, SA has been selected as the optimization algorithm for our placement tool. The remainder of this chapter will be used to explain the various aspects of our implementation of SA for analog circuit level placement.

4.5 Simulated Annealing for Analog Performance Driven Placement

The mapping of the analog placement problem into simulated annealing is accomplished as follows. The placeable entities are initially laid down in a random fashion and an initial temperature T_0 is calculated. The placer proceeds by attempting many small placement *moves* in which one or more objects are changed or relocated. After each move, a *cost function* is evaluated to determine the effect of the move on such quality measures as the estimated performance and area. If the change in the cost function ΔC is less than or equal to zero, the overall quality of the placement has improved (or is unchanged) and the new placement is retained. If ΔC is positive, the overall quality of the placement has decreased. These uphill moves are accepted with a probability based on the Metropolis relation, that is $P[uphill] \sim e^{\frac{-\Delta C}{T}}$. If the move is rejected, the placement is returned to its previous state. After a sufficiently large number of successful moves, a decision is made that the system is in thermal equilibrium. At this point, a new lower temperature is calculated and the process begins anew. Eventually the placement no longer improves with new moves at a lower temperature and the layout is said to be *frozen* and the optimization complete. This process is outlined in Fig. 4.4. To be able to apply the simulated annealing algorithm to the analog performance driven placement problem, we have to define the configurations of the problem, the move generation mechanism and the cost function. Defining the configuration of the problem comes down to finding a suitable representation of the evolving placement and all objects related to the placement problem. This will be discussed in section 4.5.1. The generation mechanism is implemented by a set of moves which is used to perturb a placement and is discussed in section 4.7. The cost function used to evaluate the quality of a placement solution is the subject of section 4.8.

4.5.1 Placement Representation

In simulated annealing based placement, two different approaches can be used to represent placement solutions : the flat style and the slicing style. The choice of a particular placement style has implications for the move set and for the cost function of the placement algorithm.

```
Anneal Placement {
    Calculate initial temperature T0;
    Generate initial placement;
    Evaluate cost function C;
    until (placement frozen) {
        until (equilibrium reached at current T) {
            Generate random move;
            Evaluate change in cost function△C ;
            if (△C < 0} {
                    accept move;
            }
            else if ( Metropolis criterion ) {
                    accept move;
            }
            else {
                    reject move;
                    restore previous state;
            }
        }
        Decrement temperature T;
    }
}
```

Figure 4.4: The simulated annealing placement algorithm

4.5.1.1   Flat Representation

In the flat placement style, also called Gellat-Jepsen style [Jeps 84], a placement is represented by specifying the absolute coordinates of all devices. An annealer manipulates the placement by shifting the coordinates of the devices. Since there are no restrictions on the positions of the devices with respect to each other, devices can overlap in intermediate placement solutions. This illegal overlap must be driven to zero in the final result by inserting an overlap penalty term in the cost function.

4.5.1.2   Slicing Style Representation

In the slicing style [Wong 86], a placement is specified by the relative positions of all devices with respect to each other. This is done by using a slicing structure to represent a placement solution. Such a slicing structure is obtained by repeatedly partitioning the layout area into horizontal and/or vertical slices, as shown in Fig. 4.5(a). The layout area is partitioned into as many partitions as there are devices in the circuit, and each device is assigned to one partition. A slicing tree is a convenient way to represent such a slicing partition (Fig. 4.5(b)). In this tree, ∗ and + are two-operand operators, symbolizing a vertical and a horizontal cut, respectively. An annealer can perform object moves by operating directly on the slicing tree, e.g. by interchanging

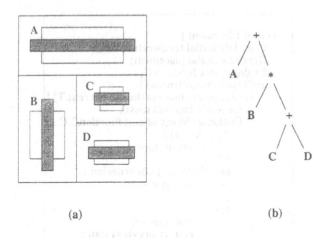

(a)   (b)

Figure 4.5: Slicing style placement representation

two neighboring operands or operators.

4.5.1.3 Discussion

Both types of placement representation have their advantages and their drawbacks.

As pointed out in section 4.2, symmetric placement is very important for high-performance analog circuit layout. The use of a flat placement representation allows the annealer to operate directly on the absolute coordinates of the devices. This makes it possible to implement the important symmetry and self-symmetry constraints directly in the move set, as will be discussed in section 4.7. Slicing style placement tools have to implement global symmetry constraints in the cost-function through the use of virtual symmetry axes [Malavasi 91], which is a less efficient solution.

Knowledge of absolute device coordinates is also necessary to accurately estimate the values of layout parasitics and to calculate the resulting performance degradation. These absolute coordinates are readily available when a flat placement style is used. If a slicing style representation is used, the relative position of the devices have to be mapped to absolute coordinates before the performance degradation can be computed. This is again a waste of CPU time and a less elegant solution.

The main advantage of a slicing style placement tool becomes clear when it is used in a digital layout system that employs channel routing. In that case, the slicing tree structure defines a channel routing order that is guaranteed to be conflict-free. In addition to this, there is never a wire space problem since the wiring channels can easily be adjusted to accommodate the required amount of routing space. However, these features offer little advantage when used in the context of analog layout. It will be shown in chapter 5 that channel routing is a poor choice for analog layout. Hence, an area router must be used and the main advantages of the slicing style over the flat placement style are lost.

The disadvantage of flat placement representations is that devices are allowed to overlap in intermediate solutions and that the placer has to drive this illegal overlap to zero during annealing. Slicing style placers avoid the overlap problem which makes them more efficient. However, this disadvantage can be turned into an advantage if it is used to explore beneficial device merging possibilities.

The last drawback of slicing style placers is that they restrict the set of reachable layout topologies. The use of a flat placement representation allows to explore *all* placement configurations, not just the ones that can be represented by slicing structures, which results in denser layouts, especially for placements with devices of widely varying sizes.

All these factors combine to make flat placement representations the better choice for analog placement.

4.5.2 Device Representation

The placement interface of a device is a geometrical abstraction of the module, that contains all information necessary to evaluate the placement cost function. For individual devices, the interface is generated by the device generator. For imported structures, the interface has to be generated based on the layout.

To evaluate the overlap penalty term in the placement cost function, the overlap between all pairs of devices has to known. To perform this computation, every device in the placement needs to have information about its boundary. For simple rectangular devices, this boundary can be represented by a box. For general rectilinear devices, the boundary is given by a rectilinear polygon. However, the overlap computations that have to be done during placement optimization are more easily performed by considering the rectilinear polygon as being composed of several rectangles. The overlap computation between two rectilinear devices is then performed by computing the overlap between every rectangle of the first device and every rectangle of the second device. If the first device is represented by n rectangles and the second by m rectangles, $\frac{1}{2} \times n \times m$ rectangle overlaps have to be computed. Since this overlap computation has to be done over and over again during the course of the annealing process, it is very important to keep the number of rectangles used to represent the geometry of a device as small as possible. Fig. 4.6 shows two different ways of covering a rectilinear polygon with rectangles. For overlap computation, cover (b) is the best. How to derive this cover is discussed next. A set T of rectangles is a cover of a polygon P if and only if P is the union of the rectangles of T. A cover is a partition of P if all rectangles are disjoint. The minimal overlapping cover (MOC) of a rectilinear polygon P is a cover of P that has the fewest number of rectangles. For overlap computations, the most efficient representation of a rectilinear polygon is its MOC. Finding the MOC of general non-convex rectilinear polygons is NP-hard. In [Wu 90], several approximation algorithms with varying complexities are reported. We have implemented an approximation algorithm with complexity $O(n^4)$, where n is the number of vertices of the rectilinear polygon. Since the number of vertices is generally limited in the problems we are dealing with, and the covering algorithm has to be executed only once for each non-rectangular module, this complexity is acceptable. Note that for a rectangular module, the minimum overlapping cover is equal to the bounding box of

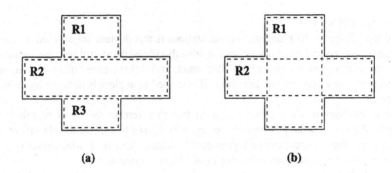

Figure 4.6: Two different covers of a rectilinear polygon :
  Cover (a) : a partition consisting of 3 rectangles,
  Cover (b) : minimum overlapping cover (MOC) consisting of only two rectangles.

the module.

The representation discussed above is a 'black box' approach : only the boundary of a device is considered. This is sufficient for placement of general rectilinear cells. However, if layout geometry sharing (see chapter 3) is considered, the geometry of each device terminal has to be taken into account. In this case, the overlap computation between two devices has to be performed by computing overlap between every terminal of the first device and every terminal of the second device and the distinction between legal and illegal overlap has to made.

Overlap between two module terminals is *legal* if and only if [Cohn 91]:

- the terminals are connected to the same net,
- the terminals belong to the same layer,
- the devices of which they are part have the same bulk potential.

In all other cases, the overlap between device terminals is illegal. Legal overlap results in a reduction of diffusion capacitance and hence in a reduction of performance degradation. Illegal overlap results in DRC errors and/or circuit topology changes and must be driven to zero. A terminal can have an arbitrary rectilinear shape. Therefore, we need to cover it with rectangles (see Fig. 4.7). Since the computation of the terminal capacitance is based on the area of the rectangles in the cover, we need a set of disjoint rectangles. Use of a minimum overlapping cover would result in an overestimation of the terminal capacitance.

The geometry of each terminal is also needed to estimate the topology of the net that is connected to it. In digital placement tools, a device terminal is often represented by a point, and sometimes it is assumed that all terminal points coincide with the center of the device. For analog circuit level layout, this simplification is unacceptable. To determine accurate estimates of the interconnect parasitics, the exact geometry of each terminal has to be considered.

For the evaluation of thermally induced performance degradation, we also need the thermal profile of a device. The thermal profile of a device is a matrix representation of the temperature distribution caused by the device in its neighborhood. This profile is computed during initialization of the device, based on its power dissipation, its geometry and the thermal profile of a unit source. The thermal distribution of an intermediate placement can be evaluated by superposition of the thermal profiles of the individual modules. This will be discussed in detail in section 4.9.3.

All these considerations lead to the following representation for each device :

- **minimum overlapping cover** : a set of rectangles covering the complete layout of the device.

- **cover bounding box** : the bounding box of the minimum overlapping cover of the module layout. This box is used to speed up overlap computation. Overlap between the rectangles of the MOC's of two devices is only checked if their cover bounding boxes overlap.

- **bulk** : the net to which the bulk of the device is connected.

- **thermal profile** : the temperature distribution caused by the device.

For each terminal of the device (see Fig. 4.7):

- **terminal partition** : a set of disjoint rectangles covering the complete layout of the device terminal.

- **terminal bounding box** : the bounding box of the terminal partition of a device terminal. This box is used to speed up overlap computation in a similar way as the cover bounding box of a device.

- **layer** : the layer of the terminal.

- **net** : a pointer to the net that is connected to the terminal.

4.5.3   Interconnect Area Estimation

An important consequence of separating the placement and routing steps in automatic layout synthesis is that the placement algorithm is responsible for allocating the routing area. Estimating the correct amount of interconnect area is crucial for the overall layout process. Failure to allocate sufficient interconnect area results in unroutable placements. Allocating too much routing area means loss of density.

In our placement tool, interconnect area between the devices is allocated by appending a border around the contours of each device. This is done by expanding the edges of each rectangle of the minimum overlapping cover of a device outwards in accordance with the expected interconnect-area requirements. The amount of interconnect area can be different for each direction. In Fig. 4.8, this procedure is illustrated for an L-shaped device. Δint_l, Δint_b, Δint_r and Δint_t are the expansion distances in the left, right, up and down directions respectively.

The interconnect distances Δint_x are computed as a sum of two terms :

$$\Delta int_x = \Delta int_{x,st} + \Delta int_{x,d},\qquad(4.4)$$

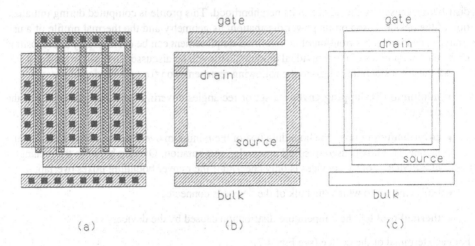

Figure 4.7: Terminal representation :
        (a) layout,
        (b) terminal partitions,
        (c) terminal bounding boxes

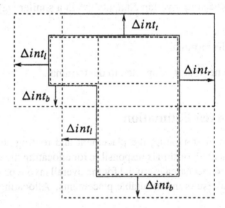

Figure 4.8: Expanding a device to allocate interconnect area

where the subscript x denotes one of the four directions l,b,r or t. $\Delta int_{x,st}$ is the static and $\Delta int_{x,d}$ is the dynamic component of the estimated interconnect area.

The static term $\Delta int_{x,st}$ is used to allocate interconnect area for nets that are connected to the device itself. This term only depends on the device itself. The static interconnect area for a direction x is computed as the sum of the widths of the wires that connect to device terminals lying on the x side edges of the device :

$$\Delta int_{x,st} = \sum_{i=1}^{T_x} W_i, \qquad (4.5)$$

where T_x is the number of terminals on the x side of the device and W_i the width of the wire connecting to terminal i. W_i is computed based on the current flowing through terminal i.

The static term $\Delta int_{x,st}$ is independent of the position of the device in the placement and remains constant during placement optimization. The dynamic component $\Delta int_{x,d}$ reserves routing space for wires that are not connected to the device and that have to pass along its edges. Its value depends on the configuration of the device terminals in its neighborhood and therefore is strongly dependent on the placement. $\Delta int_{x,d}$ has to be computed for each intermediate placement, based on the estimated topologies of the nets. We will discuss this term in section 4.10.

4.6 Handling Analog Constraints in Simulated Annealing

Up to now we have discussed the basics of simulated annealing based placement. An important question that can be raised at this point is how and where to implement the different analog constraints that have been identified in section 4.2. The effect of these constraints is that they make a subset of the complete set of possible placements illegal. For instance, a placement where the performance degradation induced by the combined effect of all interconnect parasitics exceeds one or more performance specifications violates a performance constraint and is an illegal solution. Another example is a placement where two matching devices have different orientations. Two different approaches can be taken to avoid these illegal solutions.

The first approach is to design the move set such that unacceptable placements can never be reached. In the case of two matching transistors for instance, this can be accomplished by giving them equal orientations in the initial solution and always moving the orientations simultaneously, such that the orientations are equal at all times during the annealing process and therefore also in the final solution. The advantage of this approach is that there is no CPU time wasted evaluating placements which are considered unacceptable. In addition to this, the constraints are guaranteed to be met by construction. The price that has to be paid for this is a more complicated move set.

The second approach is to put a penalty on constraint violations in the cost function. For the constraint to be met, this penalty term has to be driven to zero by the annealing mechanism. For the case of two matching transistors, this penalty term can be set to zero if their orientations are equal and to some non-zero value if they are different. An important consequence of this approach is that the constraint is no longer guaranteed to be met by construction. The simulated annealing mechanism tries to minimize the overall value of the cost function, which does not necessarily mean that every single term will be driven to zero. Implementing a constraint in the cost function implies that it will be traded off against other competing constraints and that it will be driven to zero only if this gives the best overall result.

A conclusion that can be drawn from the preceding discussion is that *hard* constraints, i.e. constraints which absolutely must be met, are best implemented as restrictions in the move set.

Unfortunately, some hard constraints are difficult to maintain by construction. If this is the case, they must be implemented as penalty terms in the cost function and special care must be taken that they are actually driven to zero in the final result (for instance by giving them large weights). The different constraints are implemented as follows in the LAYLA placement tool.

- Symmetry Constraints

  Symmetry is considered to be an absolute constraint. If the user specifies a number of devices as being symmetric and/or self-symmetric, the devices have to be symmetric in the resulting placement. Consequently, symmetry constraints are handled as restrictions in the move set. Groups of symmetric devices are moved simultaneously such that their symmetry is preserved at all times during the optimization and also in the final result.

- Matching Constraints

  If a user specifies a group of devices as a matching group, this will have two effects. First, they will have equal orientations and variants in the final placement. Second, the distance between them will be optimized in view of its influence on the performance of the circuit. The first requirement is implemented as an equal orientation/equal variant constraint in the move set. The second one is handled by including the distance between matching devices in the set of parasitic effects for which the degradation of the performance characteristics is calculated and included in the placement cost function.

  In this way, the user can specify a pair of devices as being matched without specifying the degree of matching. Matching devices are always generated identically and with identical orientations but it is up to the placement tool to determine the positions and therefore also the distance between the matched devices such that the circuit performance constraints are met. Since it is not always possible in an analog circuit layout to, at the same time, meet all symmetry requirements, put all matching devices directly next to each other and obtain a fairly compact layout, the matching degree of a pair of devices is selected in view of its influence on the performance of the circuit.

- Performance Constraints

  The most important requirement of an analog performance driven placement tool is to make sure that the performance degradation induced by the various parasitic layout effects remains within the circuit specifications. However, performance constraints can not simply be translated to restrictions on the coordinates and/or the orientations of the devices. They have to be evaluated based on an intermediate placement solution. It is therefore impossible to maintain performance constraints by construction and they have to be implemented by penalty terms in the cost function. Special care is taken to guarantee that performance constraint violations are actually driven to zero in the final result. This will be explained in section 4.8.

- Geometrical Constraints

  Geometrical constraints can be considered as hard constraints or as soft constraints. It is up to the user to make the distinction. For instance, if the height of a placement has to be smaller than a certain value to make it fit into a system level placement, the minimum height constraint is a hard constraint and is implemented in the move set. Another situation where hard geometrical constraints are imposed is when the resulting placement will be used in a standard cell layout assembly system : in that case the height and the power supply terminal positions are fixed. In cases where the geometrical constraints are specified as optimization targets they are implemented in the cost function. An example of this type is the target aspect ratio that can be specified for a placement.

- Overlap Constraint

  The overlap constraint is a hard constraint : if the final placement contains illegal overlap, it can not be used. However, it is not implemented as a restriction in the move set. Implementing overlap constraints in the move set would imply that placements with overlapping modules are never considered. A restriction like that would make it impossible to detect those situations where overlap is beneficial, for density as well as for performance reasons. Therefore, each overlap situation has to be considered individually and overlap is best dealt with in the cost function.

An important consequence of the strategy outlined above is that the annealing cost function contains penalty terms that have to be driven to zero in order to obtain a useful result. To make sure that they are actually driven to zero, special techniques must be used, as will be explained in section 4.8.

4.7   Move Set

As explained in section 4.4.6, the use of simulated annealing presupposes a generation mechanism to generate a new configuration based on the current one. In the context of analog placement, the generation mechanism is implemented as a set of moves that operate on the positions, the orientations and the variants of the devices. As described in the previous section, the move set that is used in the placement annealer has been designed to maintain symmetry and equal orientation/variant constraints by construction. As a consequence, devices which are involved in a matching and/or symmetry group can not be moved independently. If their orientation, variant or position is altered, this has to be done simultaneously with the orientation, variant or position of their symmetric or matching counterpart(s) and such that the constraint in which they are involved is not violated.

To handle these constraints in an elegant way, we have introduced the concept of a *group*. A group is a collection of one or more devices, on which one or more move types can be executed without the risk of violating a constraint. Three independent variable types are associated with each device : position, orientation and variant. A group controls one independent variable type of a collection of one or more devices simultaneously. Groups are constructed during the initialization of the program, based on the constraints specified in the input netlist. During execution of the annealing algorithm, the positions, orientations and variants of the devices are never manipulated directly but always through the groups to which they belong. A device always belongs

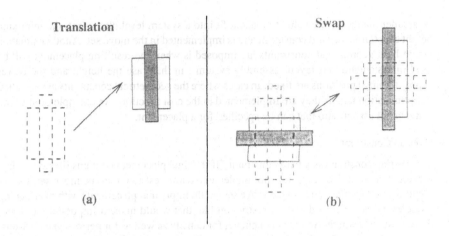

Figure 4.9: Independent relocation moves

to three groups : one to control its position, one for its orientation and one for its variant. The moves can be divided into three classes :

1. **relocation moves** change the location of one or more devices. They are executed on the *independent group* or on a *symmetry group*.

2. **reorientation moves** operate on an *orientation group* and change the orientation of one or more devices.

3. **reshaping moves** operate on a *shape group* and change the variant of one or more devices.

We will now discuss the various groups and the moves that can be executed on them.

- **Independent Group** The independent group consists of all devices that are not involved in a symmetry constraint. Consequently, their positions can be altered independently of the positions of all other devices in the circuit. There is only one independent group. Two types of relocation moves can be executed on the independent group (see Fig. 4.9):

  - **Independent Translation** (see Fig. 4.9(a)): one device of the independent group is selected at random and its center position is translated to a randomly chosen new position.

  - **Independent Swap** (see Fig. 4.9(b)): two devices of the independent group are chosen at random and their center coordinates are interchanged.

- **Symmetry Group** A symmetry group consists of all devices which are symmetric with respect to the same axis. The devices are stored as a collection of *symmetric units*. There are two types of symmetric units : *couples* and *self-symmetrics*. Two devices which have

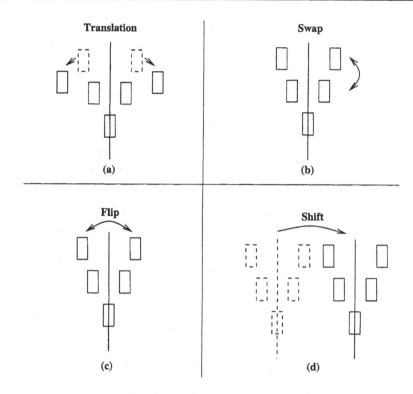

Figure 4.10: Symmetric relocation moves.

to be placed symmetrically with respect to the axis form a couple. A self-symmetric is a device that has to be placed on the symmetry axis. There can be more than one symmetry group in a placement, depending on the input netlist. A symmetry is of the X-type (Y-type) if its defining axis is parallel to the X-axis (Y-axis). Four types of relocation moves are possible for a symmetry group (see Fig. 4.10):

- **Symmetric Translation** (see Fig. 4.10(a)): a symmetric unit is randomly chosen and relocated to a new, random position. If a couple is selected, the positions of the two devices are changed, such that their symmetry is preserved. If a self-symmetric is chosen, it is shifted on the axis.

- **Symmetric Swap** (see Fig. 4.10(b)): two symmetric units are selected and their co-ordinates in the direction of the symmetry axis are interchanged. Their coordinates perpendicular to the axis remain the same.

- **Symmetric Flip** (see Fig. 4.10(c)): a couple is randomly chosen and the center coor-dinates of the two devices are interchanged.

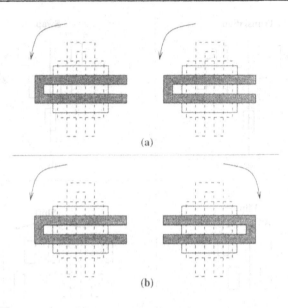

Figure 4.11: Reorientation moves

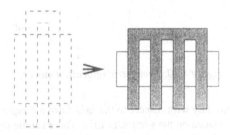

Figure 4.12: Reshaping move

– **Symmetric Shift** (see Fig. 4.10(d)): the axis of the symmetry group is shifted by a random amount in a direction perpendicular to its own direction. During this operation, the position of the symmetric units relative to the axis remains the same, but their absolute position is shifted by the same amount as the axis. This move is necessary if there are multiple symmetry groups present in the circuit.

• **Orientation Group** An orientation group is a collection of one or more devices of which the orientations are related to each other. Reorientation moves change the orientation of all devices in the group simultaneously. There are two types of orientation groups : *equal orientation groups* and *mirrored orientation* groups. The orientations of devices belonging to an equal orientation group are always equal (see Fig. 4.11(a). Those of devices belonging to a mirrored orientation group are always *mirror-symmetric* with respect to an axis. In Fig. 4.11(b), the orientations of the devices are mirror-symmetric with respect to the X-axis. Equal orientation groups are used for matching devices. Mirrored orientation groups are used to maintain the mirror-symmetry of symmetric couples. As discussed in section 4.5.2, each device is represented by a set of rectangles that cover its shape and another set of rectangles for each terminal. Changing the orientation of a device involves rotating and/or mirroring each of these rectangles, which is a time-consuming operation. To speed up this move, the layout of each device is generated for all its possible orientations during initialization of the program. The reorientation moves are then actually implemented by a variant change. This only involves a switch of a pointer.

- **Shape Group** A shape group is a collection of one or more identical devices which have to be implemented with equal variants. Reshaping moves change the variant of all devices in the group simultaneously (see Fig. 4.12). Shape groups are used for matching devices and symmetric couples.

Now that we have defined the different groups and the moves that operate on them, we can describe more formally the effect of the symmetry and matching constraints that are defined in the netlist :

1. If there is no symmetry or matching constraint defined for a device :

   - The device is inserted into the independent group.
   - An orientation group is created and the device is inserted.
   - A shape group is created and the device is inserted.

2. If a number of devices are specified as matched :

   - The devices are inserted into the independent group.
   - One equal orientation group is created and all devices are inserted.
   - If the devices are identical, one shape group is created and all devices are inserted. Otherwise, they are partitioned into groups of identical devices and a shape group is created for each subset of identical devices.

3. If a number of devices are specified as symmetric :

   - A symmetry group is created and the devices, grouped into couples and self-symmetrics, are inserted.
   - For each couple, a mirrored orientation group is created and the two devices are inserted. For each selfsymmetric, an orientation group is created and the device is inserted.
   - For each couple, a shape group is created and the devices are inserted. For each selfsymmetric, a shape group is created and the device is inserted.

4.8 Cost Function

The search for an optimum placement is driven by the cost function of the simulated annealing algorithm. The cost function is designed to minimize the area of the final placement, to drive illegal device overlap to zero and to enforce aspect ratio and performance constraints. This cost function C is calculated for each intermediate placement result and is a weighted sum of 4 terms:

$$C = \alpha C_{area} + \beta C_{aspectRatio} + \gamma C_{overlap} + \delta C_{perfDegr} \tag{4.6}$$

where :

- **Area Cost :** C_{area}
  This term is designed to minimize the area of the layout and is equal to the area of the bounding box of the intermediate placement. The reason for inclusion of this term in the cost function is to improve the density of the layout and hence to produce more cost-effective layouts.

- **Aspect Ratio Cost :** $C_{aspectRatio}$
  This term is used to drive the aspect ratio of the final placement to the value specified by the user. Its value is given by the deviation of the aspect ratio of the intermediate placement to the aspect ratio specified by the user :

$$C_{aspectRatio} = |aspectRatio - desiredAspectRatio| \tag{4.7}$$

- **Overlap Cost :** $C_{overlap}$
  Since we are using a flat placement representation, devices are allowed to overlap during the course of the simulated annealing algorithm. The overlap term in the cost function is used to drive the overlap between devices to zero in the final result. The term is given by the total amount of illegal overlap present in the intermediate placement :

$$C_{overlap} = \sum_{i=1}^{n} \sum_{j=i+1}^{n} areaOverlap_{ij} \tag{4.8}$$

where n is the number of devices in the circuit and $areaOverlap_{ij}$ the overlap area between device i and j.

- **Performance Cost :** $C_{perfDegr}$
  This term is used to keep the performance degradation induced by the various parasitic layout effects within user-specified limits. Its value is zero if all performance characteristics are within their specifications and proportional to the amount of violation if they are not. The computation of this performance cost term is crucial for our performance driven placement approach and will be explained in detail in the next section.

The weighting coefficients α, β, γ and δ are used to dynamically adjust the relative importance of each term during the course of the optimization. In the earlier stages of the optimization, when the general configuration of the placement is determined, the aspect ratio and performance terms have to dominate the cost function. Towards the end of the optimization, when the final positions of the devices are optimized without major configuration changes, the relative weight of the overlap term has to be increased to make sure that no illegal overlap is present in the final solution. To achieve this, the weighting coefficients are varied between a minimum and maximum value. After each inner loop, the relative weight of the performance and aspect ratio terms is linearly decreased from the maximum towards the minimum value, while the weight of the overlap term is increased from the minimum to the maximum value.

4.9 Estimating Performance Degradation

Consider a circuit for which N_p performance specifications have been given :

$$P_i \in [P_{min,i}, P_{max,i}] \qquad i = 1, \ldots, N_p. \tag{4.9}$$

Using the technique described in section 2.2, these performance specifications can be transformed into N_p constraints on the layout induced performance degradation :

$$\Delta P_{lay,i} \in [\Delta P_{lay,i}^{min}, \Delta P_{lay,i}^{max}] \qquad i = 1, \ldots, N_p, \tag{4.10}$$

where $\Delta P_{lay,i}$ is the layout induced performance degradation for performance characteristic P_i. The limits $\Delta P_{lay,i}^{min}$ and $\Delta P_{lay,i}^{max}$ can be calculated using equations (2.6) and (2.7) respectively.

A penalty term $C_{perfDegr,i}$ is associated with each performance characteristic P_i. If (4.10) is satisfied for P_i, $C_{perfDegr,i}$ is set to zero. If not, its value is proportional to the amount of violation :

$$C_{perfDegr} = \begin{cases} \Delta P_{lay,i}^{min} - \Delta P_{lay,i} & \text{if } \Delta P_{lay,i} < \Delta P_{lay,i}^{min}, \\ 0 & \text{if } \Delta P_{lay,i}^{min} \leq \Delta P_{lay,i} \leq \Delta P_{lay,i}^{max}, \\ \Delta P_{lay,i} - \Delta P_{lay,i}^{max} & \text{if } \Delta P_{lay,i} > \Delta P_{lay,i}^{max}. \end{cases} \tag{4.11}$$

The total performance degradation cost term is given by the sum of the penalty terms for each performance characteristic :

$$C_{perfDegr} = \sum_{i=1}^{N_p} C_{perfDegr,i} \tag{4.12}$$

The evaluation of (4.11) for an intermediate placement solution requires the knowledge of the layout induced performance degradation $\Delta P_{lay,i}$ for each performance characteristic P_i. To compute this performance degradation we follow the direct performance driven methodology which was explained in section 2.2. Based on the geometrical information of the intermediate placement, we estimate the value of all parasitic layout effects. To compute the influence of the parasitic layout effects on the performance characteristics of the circuit, we use a linear approximation based on the performance sensitivities, which are derived once from simulations before placement starts. This methodology will now be applied to interconnect parasitics, device mismatches and thermal effects.

4.9.1   Interconnect Parasitics

This section describes the estimation of performance degradation due to interconnect parasitics. In section 4.9.1.1, we discuss the technique that is used to estimate the topology of a net during placement. In section 4.9.1.2 we describe how interconnect parasitics can be estimated based on the net topology, and how the resulting performance degradation can be evaluated.

4.9.1.1   Net Topology Estimation

A complete and accurate extraction of the interconnect parasitics of a layout requires the knowledge of the exact layout of each net. During placement, the layout of the nets is unknown and hence the interconnect parasitics can only be estimated. To make a reasonably accurate estimation of the different parasitics, a technique to estimate the geometry of a net based on the locations of its connecting terminals is needed. To model the interconnect, we only consider the parasitic resistance and capacitance (see section 2.5). To estimate the series resistance and the capacitance to ground of a net, an estimation of the total *length* of a net is needed. To make reasonable estimations of coupling capacitances between different nets, we also need the *geometry* of the nets. Net estimation techniques which have been proposed in the past have focussed on net *length* estimation. In this section, we give an overview of the most commonly used techniques and we show how the minimum spanning tree technique is used to give reasonably accurate estimations for length and geometry estimations in a computationally efficient way.

Given a net of n terminals, the following techniques can be used as net length estimators [Shah 91, Cohn 94]:

- **(a) semi-perimeter** (Fig. 4.13(a)) : this technique estimates the net-length as the length of half the perimeter of the smallest rectangle that contains the centers of all terminals. The semi-perimeter can be computed in $O(n)$.

- **(b) minimum spanning tree** (Fig. 4.13(b)) : the net-length is estimated as the sum of the length of the $n - 1$ paths of a minimum spanning tree connecting the centers of all terminals. A minimum spanning tree is the minimum length acyclic connected path that connects the centers of all terminals [Thul 92]. Several algorithms of complexity $O(n^2)$ have been proposed [Krus 56, Prim 57].

- **(c) center of mass** (Fig. 4.13(c)) : the net-length is estimated as the sum of the distance of all terminal centers to the weighted mean center of the net. Computation complexity is $O(n^2)$.

- **(d) minimum Steiner tree** (Fig. 4.13(d)) : this technique uses the sum of the path-lengths of a minimum Steiner tree as an approximation for the net-length. In a Steiner tree, a path can branch at any point along its length. A minimum Steiner tree is the shortest possible route for connecting a set of terminals. The computation of a Steiner tree is an NP-complete problem. Heuristics can be used to find a non-optimal solution with complexity ranging from $O(n log n)$ to $O(n^2)$ [Oht 86].

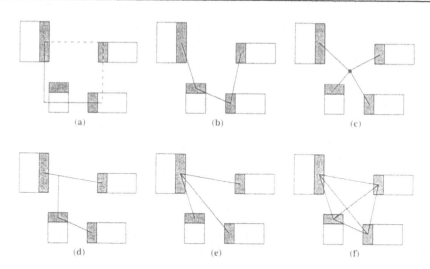

Figure 4.13: Net topology estimation techniques :
    (a) semi-perimeter
    (b) minimum spanning tree
    (c) center of mass
    (d) Steiner tree
    (e) source to sink
    (f) complete graph

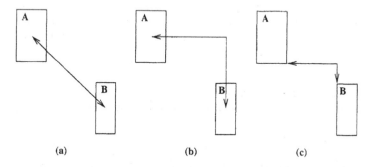

Figure 4.14: Different ways to measure the distance between terminals A and B :
    (a) Euclidean center to center
    (b) Manhattan center to center
    (c) Manhattan edge to edge

- **(e) source to sink** (Fig. 4.13(e)) : in this technique, one *source* module is connected to all other *sink* modules and the total length of these connections is used to approximate the net-length. The complexity of this approach is $O(n)$.

- **(f) complete graph** (Fig. 4.13(f)) : net-length is estimated using a complete graph connection of the centers of all terminals. The calculation complexity of this measure is $O(n^2)$.

A realistic evaluation of interconnect parasitics requires an accurate prediction of the length and the location of the nets. In addition to this, the estimator must be efficient enough to make it suitable for use in the inner loop of a simulated annealing algorithm. Given these requirements, the minimum spanning tree technique has been selected for our placement tool. To compute the minimum spanning tree, we use the algorithm presented in [Thul 92], which is based on the algorithms of Prim [Prim 57] and Kruskal [Krus 56].

Several distance metrics can be used as a basis for computing the minimum spanning tree (see Fig. 4.14). Among these the *edge to edge Manhattan distance*, which was proposed in [Cohn 94], gives the most accurate results for analog layouts and was selected for our placement tool.

4.9.1.2 Interconnect Parasitics Extraction

Using a minimum spanning tree to approximate the geometry of each net and the interconnect parasitic models described in chapter 2, the interconnect parasitics can be extracted from the placement. This is illustrated in Fig. 4.15 with a fragment of a placement with five devices and two nets.

Net n_1 connects terminals $T1, T2, T3$ and net n_2 terminals $T4, T5$. To compute the parasitics, both nets are approximated by their minimum spanning trees : the minimum spanning tree for n_1 consists of the paths $T1 \rightarrow T2$ and $T2 \rightarrow T3$, the one for n_2 consists of the path $T4 \rightarrow T5$. We define $L(TX \rightarrow TY)$ and $W(TX \rightarrow TY)$ as the length and the width of the path $TX \rightarrow TY$. $L(TX \rightarrow TY)$ can be extracted from the placement and $W(TX \rightarrow TY)$ can be computed based on the current flowing into terminals TX and TY. The following parasitics can now be computed :

- **Capacitance to ground C_x of a net n_x**
  The total capacitance to ground of net n_x is the sum of two terms:

$$C_x = C_{w,x} + C_{t,x} \tag{4.13}$$

where $C_{w,x}$ is the estimated capacitance of the wire to ground and $C_{t,x}$ is the sum of the capacitances of the terminals which are connected to the net. $C_{w,x}$ can be calculated as the total area of the net multiplied by C_{av}, the average capacitance to ground per unit area. C_{av} is computed as a weighted average of the capacitance per unit area of the different routing layers that are used in the process. Applied to nets n_1 and n_2, this gives :

$$C_{w,1} = [L(T1 \rightarrow T2)W(T1 \rightarrow T2) + L(T2 \rightarrow T3)W(T2 \rightarrow T3)] C_{av} \tag{4.14}$$

$$C_{w,2} = [L(T4 \rightarrow T5)W(T4 \rightarrow T5)] C_{av}. \tag{4.15}$$

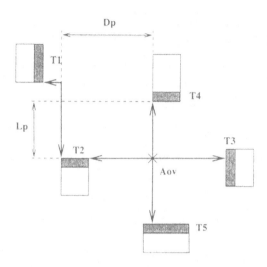

Figure 4.15: *Extraction of interconnect parasitics from an intermediate placement,*
net $n_1 = T1, T2, T3$
net $n_2 = T4, T5$

The total terminal capacitance $C_{t,x}$ can be computed as the sum of the capacitance of each connecting terminal. For nets n_1 and n_2 :

$$C_{t,1} = C_{T1} + C_{T2} + C_{T3} \tag{4.16}$$
$$C_{t,2} = C_{T4} + C_{T5}. \tag{4.17}$$

- **Coupling capacitance C_{xy} between two nets n_x and n_y**
  The coupling capacitance C_{xy} between two nodes n_x and n_y is estimated using the minimum spanning tree approximations for the two nets. Let MST_x and MST_y be the minimum spanning tree approximations of net n_x and n_y respectively. Every time a path from MST_x overlaps or runs in parallel with a path from MST_y an amount of capacitance has to be added to the coupling capacitance C_{xy}. If there are N_x (N_y) paths in MST_x (MST_y), the total coupling capacitance C_{xy} is given by :

$$C_{xy} = \sum_{i=1}^{N_x} \sum_{j=1}^{N_y} \left[L_{p,ij} C_c(D_{p,ij}) + C_{ov}(A_{ov,ij}) \right] \tag{4.18}$$

where $L_{p,ij}$ is the length that paths i and j run in parallel, $D_{p,ij}$ the distance between their parallel segments, and $A_{ov,ij}$ the overlap area between them. $C_c(d)$ gives the average coupling capacitance per unit length as a function of the distance d between two pieces and $C_{ov}(a)$ gives the average coupling capacitance per unit area as a function of the overlap area *a.* In Fig. 4.15, path $T1 \rightarrow T2$ runs in parallel with path $T4 \rightarrow T5$ over a distance L_p and a separation D_p, and path $T2 \rightarrow T3$ overlaps path $T4 \rightarrow T5$. The coupling capacitance C_{12} is thus given by :

$$C_{12} = L_p C_c(D_p) + C_{ov}(A_{ov}). \tag{4.19}$$

The overlap area A_{ov} can be computed as :

$$A_{ov} = W(T2 \rightarrow T3) \times W(T4 \rightarrow T5). \tag{4.20}$$

- **Series resistances $R_{x,i}, i = 1, ...N_t$ of net n_x**
  The series resistances $R_{x,i}, i = 1..N_t$ of the wire segments are calculated as :

$$R_{x,i} = \rho_{square,av} \frac{L}{\frac{t_x}{W_{wire,i}}}, \tag{4.21}$$

where L is the estimated total net-length and t_x the number of terminals connected to the net. $W_{wire,i}$ is the width of the ith wire segment which can be calculated from the current flowing through the ith terminal. $\rho_{square,av}$ is a weighted average of the sheet resistances of the routing layers which are used in the process.

Once the value of C_x and $R_{x,i}, i = 1..n$ for every net and $C_{x,y}$ between any two nets is known, the performance degradation $\Delta P_{j,int}$ for the performance characteristic P_j due to interconnect parasitics can be determined using the precalculated sensitivity information:

$$\Delta P_j = \sum_{k=1}^{m}(S_{C_k}^{P_j} C_k + \sum_{i=1}^{t_k} S_{R_{k,i}}^{P_j} R_{k,i} + \sum_{i=1,i \neq k}^{m} \frac{1}{2} S_{C_{ki}}^{P_j} C_{ki}) \tag{4.22}$$

where m is the number of nodes minus the ground node and t_k is the number of terminals of net k. $S_{C_k}^{P_j} = \frac{\delta P_j}{\delta C_k}$, $S_{R_{k,i}}^{P_j} = \frac{\delta P_j}{\delta R_{k,i}}$ and $S_{C_{ki}}^{P_j} = \frac{\delta P_j}{\delta C_{ki}}$ are the sensitivities of performance characteristic P_j to small changes in the parasitic capacitance C_k, the parasitic resistance $R_{k,i}$ and the coupling capacitance C_{ki}, respectively. These sensitivities are determined in advance by simulation.

4.9.2 Device Mismatch

We follow a comparable approach with respect to device mismatches. $\sigma^2(V_{T0})$ and $\sigma^2(\beta)$ for the V_{T0} and β mismatch of MOS transistors can be estimated using the following equations [Pel 89] (see also chapter 2) :

$$\sigma^2(V_{T0}) = \frac{A_{V_{T0}}^2}{WL} + S_{V_{T0}}^2 D^2 \tag{4.23}$$

$$\sigma^2(\beta) = \frac{A_\beta^2}{WL} + S_\beta^2 D^2 \tag{4.24}$$

where $A_{V_{T0}}, S_{V_{T0}}, A_\beta$ and S_β are constants depending on the process. The area of the devices WL and the distance D are known for each intermediate placement. Based on this information, the standard deviations of V_{T0} and β can be calculated with (4.23) and (4.24). Using predetermined sensitivity information, the effect on the degradation of performance characteristic P_j can be estimated as follows:

$$\Delta P_j = \sum_{k=1}^{m} (\left| S_{\Delta V_{T0,k}}^{P_j} \right| (3\sigma(V_{TO})_k) + \left| S_{\Delta \beta_k}^{P_j} \right| (3\sigma(\beta)_k) \tag{4.25}$$

where $S_{\Delta V_{T0,k}}^{P_j} = \dfrac{\delta P_j}{\delta \Delta V_{T0,k}}$ and $S_{\Delta \beta_k}^{P_j} = \dfrac{\delta P_j}{\delta \Delta \beta_k}$ are the sensitivities of performance characteristic P_j to small changes in ΔV_{T0} and $\Delta \beta$ of matching transistor pair k. The sum is taken over all m pairs of matching devices.

Equation (4.25) can be rewritten as

$$\Delta P_j = \Delta P_{j,area} + \Delta P_{j,distance} \tag{4.26}$$

$\Delta P_{j,area}$ represents the degradation of the jth performance characteristic due to area effects. This term can be computed after sizing and remains constant during placement.

$\Delta P_{j,distance}$ represents the degradation of the jth performance characteristic due to distance effects and therefore depends on the actual layout. This term can be computed as follows :

$$\Delta P_{j,distance} = \sum_{k=1}^{m} (S_{D_k}^{P_j}(D_k)) \tag{4.27}$$

where D_k represents the distance between the transistors of matching pair k and $S_{D_k}^{P_j}$ is the sensitivity of performance characteristic P_j to small variations in distance D_k. This term must be recomputed for every new placement.

4.9.3 Thermal Effects

To estimate thermally induced performance degradation for an intermediate placement solution, the temperature of each device has to be computed by combining the contributions of all other devices. In section 2.8 of chapter 2, the thermal distribution of a rectangular power source on the surface of an integrated circuit structure was computed as :

$$T(x,y) = \frac{4 Q_0''}{k_N} \sum_{n=0}^{\infty} \sum_{m=0}^{\infty} \tau_N(n,m) \cdot \frac{\sin \dfrac{n \pi w}{L_x}}{(1+\delta_{n0}) n\pi} \cdot \frac{\sin \dfrac{m \pi h}{L_y}}{(1+\delta_{m0}) m\pi} \cdot \cos \frac{n \pi x}{L_x} \cdot \cos \frac{m \pi y}{L_y} \tag{4.28}$$

Expression (4.28) can not be used for repeated thermal analysis of a placement for two reasons. First, it is too expensive to evaluate. The required number of terms in the series for a designated accuracy is directly proportional to the ratio of the chip to source size [Lee 89]. For the analysis of structures with large chip-to-source size ratios, a considerable amount of CPU time is needed for (4.28) to converge. Consequently, it is impossible for a thermally constrained placement tool to recalculate the thermal profile of each intermediate placement using (4.28). Secondly, expression (4.28) can only be used to compute the thermal profile of rectangular power sources. Since the power sources in analog circuits can have any rectilinear shape, another technique is needed to evaluate the thermal profile of a rectilinear power source.

Therefore, we have designed an efficient thermal computation scheme based on the two-dimensional Fourier series (4.28). The basic idea of the computation scheme is to reuse information as much as possible. Thermal computation is done in three steps.

4.9.3.1   Unit Source Thermal Profile Computation

Given the thermal model and the chip dimensions, the first step consists of computing a thermal model for a unit source by evaluating series (4.28) on a grid of points (x_i, y_j), $i = 0, \cdots, P - 1$, $j = 0, \cdots, Q-1$. As discussed above, the required number of terms for each evaluation point is proportional to the chip to source ratio and can be fairly high. A direct computation of series (4.28) requires a considerable amount of CPU time, typically 30 sec for a two layer thermal model. Since this summation has to be repeated for each evaluation point, the computational burden of this step is very high.

The computational complexity of evaluating (4.28) can be reduced using the Discrete Cosine Transform (DCT), a derivative of the Fast Fourier Transform (FFT) [Ghar 95a, Ghar 95b]. The DCT of a two-dimensional series k_{nm} is defined as :

$$K_{pq} = \sum_{n=0}^{P-1} \sum_{m=0}^{Q-1} k_{nm} \cos \frac{n \pi p}{P} \cos \frac{m \pi q}{Q}. \tag{4.29}$$

(4.28) evaluated to the limits $P - 1$ and $Q - 1$ can be written as :

$$T(x, y) = \sum_{n=0}^{P-1} \sum_{m=0}^{Q-1} k_{nm} \cos \frac{n \pi x}{L_x} \cdot \cos \frac{m \pi y}{L_y}, \tag{4.30}$$

with k_{nm} given by :

$$k_{nm} = \frac{4 Q_0''}{k_N} \tau_N(n, m) \cdot \frac{\sin \frac{n \pi w}{L_x}}{(1 + \delta_{n0}) n \pi} \cdot \frac{\sin \frac{m \pi h}{L_y}}{(1 + \delta_{m0}) m \pi}. \tag{4.31}$$

By comparing expressions (4.29) and (4.30), it is easy to see that the temperature of a grid of points can be determined from the DCT of the series (4.31). To determine the temperature of a point (x, y), $\frac{x}{L_x}$ and $\frac{y}{L_y}$ are expressed as integer ratios $\frac{p}{P}$ and $\frac{q}{Q}$. The temperature is then given by the element K_{pq} of the DCT of (4.31). P and Q are chosen as the number of discretizations in the x and y coordinates respectively. Once the DCT of k_{nm} is computed, it can be stored as a matrix and reused during module thermal profile computation.

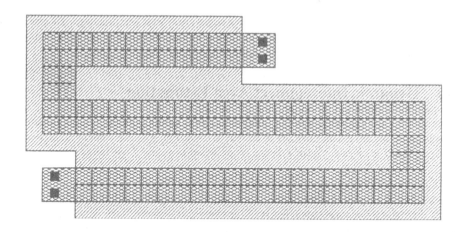

Figure 4.16: A resistor with resistive area fractured into unit areas.

4.9.3.2 Device Thermal Profile Computation

During the device initialization phase, the active surface of a variant, which can have any rectilinear shape, is discretized into a collection of unit sources. Fig. 4.16 shows an example of a resistor discretized into unit source areas. The contribution of each unit source is given by the precalculated unit source profile, shifted in position and multiplied with the actual power density of the device. As the thermal problem is linear, the thermal profile of the complete variant can be evaluated by adding up the contributions of the unit sources, and is stored in its thermal matrix. This computation has to be done only once for each variant. The resulting matrix can be reused throughout the placement.

4.9.3.3 Placement Thermal Profile Computation

During the annealing loop, the temperature distribution of an intermediate placement can be calculated by superposition of the individual devices' thermal profiles. This can be done very fast since the thermal profiles were precomputed and stored in the thermal matrices of the variants. This results in the local temperature of every device for the given placement, from which the thermally-induced performance degradation is calculated, again using predetermined sensitivity information :

$$\Delta P_j = \sum_{k=1}^{m} S_{\Delta T_k}^{P_j} \Delta T_k \tag{4.32}$$

where $S_{\Delta T_k}^{P_j} = \dfrac{\delta P_j}{\delta \Delta T_k}$ is the sensitivity of performance characteristic P_j to small differences in temperature between transistor pair k.

Studies have shown that, when the heat source edges are at least one structure thickness away from the boundaries of the rectangular structure, the thermal profiles are weakly affected by the boundaries and thus the boundaries can be assumed to extend to infinity [Lee 88].

4.10 Dynamic Interconnect Area Estimation

In section 4.5.3 we explained how the interconnect area between the devices is allocated by appending a border around the contours of each device (see Fig. 4.8. The amount of interconnect area Δint_x reserved on each side of a device is computed as a sum of two terms :

$$\Delta int_x = \Delta int_{x,st} + \Delta int_{x,d}. \tag{4.33}$$

where x is one of the possible directions l, b, r, t (left,bottom,right,top). In this equation, $\Delta int_{x,st}$ denotes the static interconnect area, reserved for wires that connect to the device itself. It was shown in section 4.5.3 that his term can be estimated based on the device terminal locations and the terminal currents.

The dynamic component of the interconnect area $\Delta int_{x,d}$ is used to reserve area for wires that do not connect to the device itself but that have to be routed around the device to connect terminals belonging to other devices in the placement. Correct estimation of this interconnect area is crucial for the overall quality of the layout. Insufficient interconnect area results in un-routable placements. Allocating too much routing area means loss of density. Unfortunately, the amount of interconnect area needed to route the layout is difficult to estimate during placement, since the exact geometry of the nets is unknown. We have explored several strategies to estimate this interconnect area and we will describe them briefly in this section.

The first approach is to make $\Delta int_{x,d}$ dependent on the location of the device in the placement. The wire density of a layout usually decreases from the center of the placement towards the edges and the interconnect area can therefore can be estimated as :

$$\Delta int_{x,d} = k_1 - k_2\sqrt{(x_e - x_c)^2 + (y_e - y_c)^2}, \tag{4.34}$$

where (x_e, y_e) is the center of the edge for which the interconnect area is estimated, and (x_c, y_c) is the center of the placement. k_1 and k_2 are two experimentally determined constants.

A second approach is to compute $\Delta int_{x,d}$ based on the estimated location of the nets. This results in the following equation for the dynamic interconnect area :

$$\Delta int_{x,d} = \sum_{i=1}^{N} \left(k_{1,i} - k_{2,i}\sqrt{(x_e - x_{c,i})^2 + (y_e - y_{c,i})^2} \right). \tag{4.35}$$

In this equation, N is the number of nets in the circuit, (x_e, y_e) is the center of the edge for which the interconnect area is estimated, and $(x_{c,i}, y_{c,i})$ is the center of the bounding box of the terminals that connect to net i. $k_{1,i}$ and $k_{2,i}$ are constants depending on the estimated wirewidth for net i, which can be computed based on the current flowing through the net. Although the computational cost of the second approach is somewhat higher than that of the first, the results obtained with this interconnect area estimation technique are consistently and significantly better.

4.11 Annealing Schedule

Simulated annealing based placement can be viewed as an algorithm that continuously attempts to transform the current placement into one of its neighbors. The neighbors are defined as the placements that can be reached from the current one by applying one of the moves described in section 4.7. Mathematically, this mechanism can be described by means of a *Markov chain*: a sequence of trials, where the outcome of each trial depends only on the outcome of the previous one [Pap 91]. A Markov chain is described by means of a set of conditional probabilities $P_{ij}(k - 1, k)$ for each pair of outcomes (i, j). $P_{ij}(k - 1, k)$ is the probability that the outcome of trial k is j, given that the outcome of trial $k - 1$ is i. If the conditional probabilities do not depend on k, the Markov chain is called *homogeneous*, otherwise it is called *inhomogeneous*. In the case of simulated annealing, the conditional probabilities depend on the value of the temperature T. Thus, if T is kept constant, the corresponding Markov chain is homogeneous and an $R \times R$ transition matrix $P = P(T)$ can be defined as :

$$
P_{ij}(T) = \begin{cases} G_{ij}(T)A_{ij}(T) & i \neq j, \\ 1 - \sum_{l=1,l\neq i} RG_{il}(T)A_{il}(T) & i = j \end{cases} \tag{4.36}
$$

where R is the number of possible placement configurations. Each transition probability is defined as the product of a generation probability $G_{ij}(T)$ and an acceptance probability $A_{ij}(T)$.

The simulated annealing algorithm used in our placement tool is of the homogeneous type [Laar 87] : it is a sequence of homogeneous Markov chains, each generated at a fixed value of T, and T is decreased in between subsequent Markov chains. It is shown in [Laar 87] that a homogeneous simulated annealing algorithm obtains a global optimum if

1. each individual Markov chain is of infinite length

2. certain conditions on the matrices $A(T_l)$ and $G(T_l)$ are satisfied;

3. $lim_{l\to\infty} T_l = 0$,

where T_l is the temperature of the l-th Markov chain.

In a practical implementation of the algorithm, this asymptotic convergence can only be approximated. The number of transitions for each temperature T_l must be finite, and $lim_{l\to\infty} T_l = 0$ can only be approximated in a finite number of temperatures T_l. Therefore, each implementation of a simulated annealing algorithm involves a trade-off between speed of execution and quality of the final solution. This speed/quality trade-off is influenced by the choice of a number of parameters which are together referred to as the *cooling schedule* of the algorithm. The cooling schedule used in our placement tool will be discussed next.

The first parameter that plays a role is the initial temperature T_0. The value of T_0 is crucial for the efficiency of the algorithm. If T_0 is too high, too much CPU time is wasted exploring the configuration space in the initial phase of the algorithm. If T_0 is too low, there is a risk of getting stuck in a local minimum. In our algorithm, the initial temperature is chosen such that the average increase in cost $\overline{\Delta C}^{+}$ is accepted with a certain probability P_0 at T_0 [Otten 84]. To

achieve this, a number of random moves are executed at the start of the algorithm and the value
of $\overline{\Delta C^+}$ is measured. T_0 is then solved from

$$P_0 = exp(\frac{\overline{\Delta C^+}}{T_0}).$$ (4.37)

This leads to the following value for T_0 :

$$T_0 = \frac{\overline{\Delta C^+}}{ln(P_0)}.$$ (4.38)

The default value for P_0 in our algorithm is 0.6. This default value can be overruled by the user.

The second important parameter is final value of the temperature, T_f. T_f is determined by
the *stopping criterion* of the algorithm. In our implementation, T_f is selected by terminating the
execution of the algorithm if the cost of the last configurations of consecutive Markov chains are
within a specified interval, for a number of chains. The width of the interval and the number of
chains can be specified by the user.

The third part of the cooling schedule is the selection of the length L_k of each Markov chain k
and the transformation rule for changing T_k into T_{k+1}. These two decisions are related through the
concept of *the stationary distribution* of a Markov chain. The stationary distribution of a Markov
chain is the probability distribution of the configurations after an infinite number of transitions.
This distribution is characterized by a vector $\mathbf{q}$, whose i-th component gives the probability of
the system to be in state i after an infinite number of transitions. For simulated annealing, $\mathbf{q}$
depends on the temperature T_k and is given by [Laar 87] :

$$q_i(T_k) = \frac{exp(\frac{-(C(i)-C_{opt})}{T_k})}{\sum_{j=1}^{R} exp(\frac{-(C(i)-C_{opt})}{T_k})},$$ (4.39)

where $C(i)$ is the cost of the i-th configuration, C_{opt} is the optimal cost and R is the number
of possible configurations. The Markov chain at T_k can be stopped if the Markov chain is in
quasi-equilibrium, i.e. if the probability distribution of the configurations is "close enough" to
the stationary distribution (4.39). Determining a rule for the length L_k of a Markov chain at temperature T_k comes down to defining the exact meaning of "close enough" and hence to determine
when the chain is in quasi-equilibrium. In our cooling schedule we use the criterion proposed
in [Cath 88]. This quasi-equilibrium rule is based on the convergence of the denominator of
(4.39). While building the chain, an estimate of this denominator is gradually updated with the
contribution of the configurations which are accessed up to that moment. The chain is stopped
when the new contributions are not changing the average value anymore.

The rule to update the temperature T_k to the next temperature T_{k+1} is related to the length of
the Markov chain at temperature T_k. The ratio between the old and the new temperature will be
denoted as α :

$$T_{k+1} = \alpha T_k,$$ (4.40)

performance				
Performance	Spec	Plac 1	Plac 2	Unit
offset voltage	< 5	3.7	6.9	*mV*
delay	< 5	2.8	5.4	*nsec*

Table 4.1: Comparator : performance characteristics

where α is varied between 0.95 and 0.8. A long Markov chain length at temperature T_k can be seen as an indication that the simulated annealing algorithm is in a critical region and hence that a high value of α should be used. A short Markov chain length indicates a less critical region and justifies the use of a smaller α value. The details of this algorithm can be found in [Cath 88].

4.12 Experimental Results

The algorithm described in this chapter has been implemented using the C^{++} language in the UNIX environment and has been integrated in a complete synthesis environment for analog circuits [Gielen 95a]. The program was tested on a number of analog circuits. Three of them are presented in this section.

4.12.1 Comparator

The first example is a high-speed CMOS comparator [Steyaert 93]. The circuit is used in a CMOS A/D converter and its performance is a limiting factor for the performance of the overall A/D converter. The specifications imposed upon the circuit are a propagation delay of less than $5nsec$ and an offset voltage of less than $5mV$. The circuit schematic is shown in Fig. 4.17. To demonstrate the effectiveness of our direct performance-driven approach we have generated two placements for this circuit. Placement 1 (see Fig. 4.18) was generated with the presented performance-driven placement tool, while placement 2 (see Fig. 4.19) was generated in the traditional way, with the same placement tool but with the performance-driven mechanism disabled. It can be seen from Table 4.1 that the simulated performance of placement 1 is significantly better than that of placement 2. Placement 1 has both performance characteristics within the user-specified ranges, while for placement 2 both specifications are violated. The optimized distances between the matching transistor pairs together with the resulting offset voltage degradation due to distance effects are shown in Table 4.2 for placement 1. The nominal values are the values obtained after sizing of the circuit without parasitic layout effects (no parasitic node capacitances and no mismatch). It can be seen that the performance-driven algorithm selectively minimizes the distances for the most sensitive transistor pairs, which results in a lower offset voltage. CPU times were 106 and 93 seconds for placement 1 and 2, respectively, on a SUN SPARC 10 workstation, which means that the performance-driven mechanism significantly improves the circuit performance at only a small increase in CPU time.

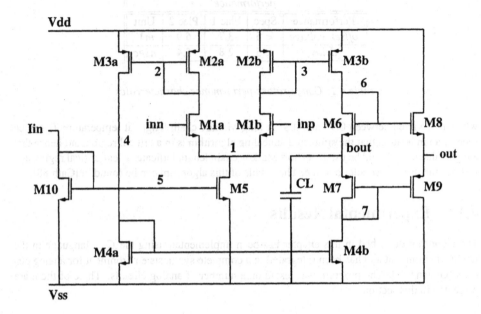

Figure 4.17: Comparator : schematic.

comparator offset voltage degradation			
Transistor Pair	Distance	Sensitivity	Degradation
	μm	$\frac{\mu V}{\mu m}$	mV
$M1 - M2$	60	12	.720
$M3 - M5$	70	2.9	.203
$M4 - M6$	70	2.9	.203
$M7 - M8$	118	.2	.024
$M11 - M12$	119	1.93	.230
$M13 - M14$	38	3.8	.145
$M15 - M16$	40	3	.120
total			1.645

Table 4.2: Comparator : offset voltage

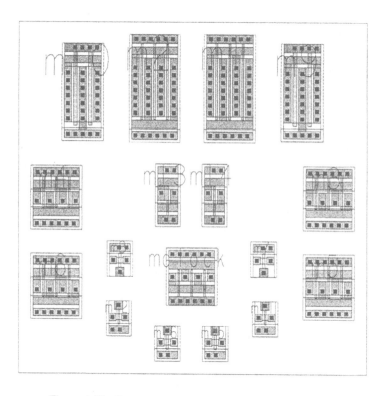

Figure 4.18: Comparator : performance driven placement.

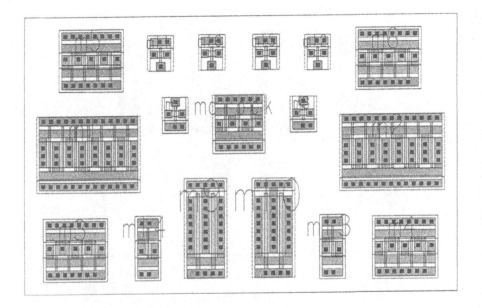

Figure 4.19: Comparator : not performance driven placement.

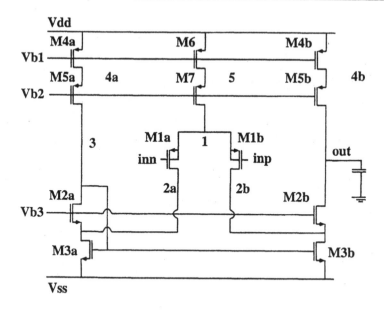

Figure 4.20: Opamp1 : schematic.

4.12.2 Opamp1

The second example is a high-speed CMOS operational amplifier [Fisher 87]. The schematic of the circuit is shown is Fig. 4.20. The placement that was generated for this circuit is shown in Fig. 4.21. The symmetry axis of the circuit is clearly visible in the placement. This circuit contains some very large transistors which have been split into parallel parts for reasons explained in chapter 3. As shown in Table 4.3 all performance characteristics for this circuit are within the specifications. The nominal values are the values obtained after sizing of the circuit without parasitic layout effects (no parasitic wire capacitances and no mismatches). All specifications were met in one pass of the program (CPU time 83 seconds). Note that the remaining performance margins after placement are needed for the subsequent routing phase (some parasitics may be different than estimated during placement).

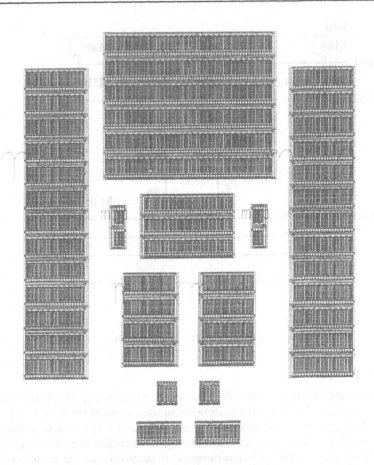

Figure 4.21: Opamp1 : placement.

opamp performance				
Performance	Specification	Nominal Value	After Placement	Unit
GBW	> 225	228	226	MHz
A_v	> 60	67	67	dB
PM	> 60	61	60.2	deg
slew − rate	> 150	163	163	$V/\mu sec$
Voffset	< 5	4.5	4.8	mV

Table 4.3: Opamp1 : performance characteristics obtained after placement.

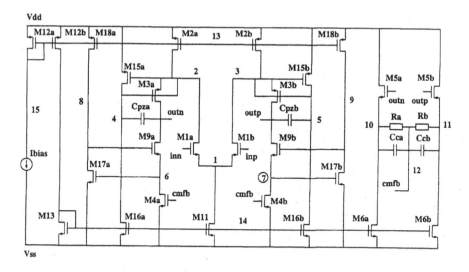

Figure 4.22: Opamp2 : schematic.

opamp performance				
Performance	Specification	Nominal Value	After Placement	Unit
A_v	> 100	107	104	dB
GBW	> 200	205	202	MHz
PM	> 70	77	74	deg
CMRR@10Hz	> 70	∞	78	dB
PSRR@10Hz	> 80	∞	86	dB

Table 4.4: Opamp2 : performance characteristics.

4.12.3 Opamp2

As a second example, a fully differential CMOS operational amplifier [Peeters 93] (see Fig. 4.22) was used to test the efficiency of the algorithm for larger circuits. The placement of the opamp is shown in Fig. 4.23. Note the clear symmetry axis in this fully differential circuit. The circuit specifications together with the obtained performances after sizing (without parasitics) and after placement are given in Table 4.4. The degradation of all performances clearly remains within the specified margins. This placement required a CPU time of 163 seconds (less than 3 minutes) on a SUN SPARC 10 workstation.

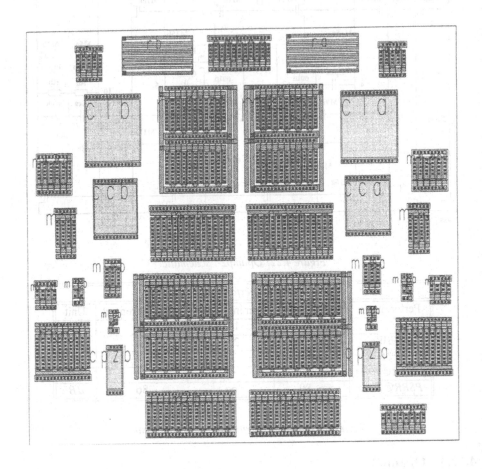

Figure 4.23: Opamp2 : placement.

Performance	Spec	Nominal	after Plac.
UGBW (MHz)	> 7	7.14	7.13
PM (^{o})	> 70	70.6	70.3
Offset Voltage (mV)	< 3	1.3	2.3

Table 4.5: Performance characteristics of the class A/B operational amplifier

opamp offset voltage			
Trans. Pair	Sens.	Temp. Diff.	Degr.
	$\frac{mV}{deg}$	deg	mV
$M1a - M1b$	1.6	0.153	0.245
$M8a - M8b$	3.7	0.084	0.319
$M13a - M13b$	6.0	0.074	0.448
total			1.012

Table 4.6: Opamp3 : offset voltage

4.12.4   Opamp3

Finally, we will illustrate the thermal capabilities of the tool with the class AB operational amplifier shown in Fig. 4.24. The power consumption of the circuit is $140mW$, of which most is dissipated in the output stage $M14, M15, M16, M17$. An offset voltage of less than 5 mV is specified.

During circuit analysis, a number of simulations with different device temperatures were done to determine the sensitivity of the offset voltage of the opamp to temperature differences between the matching transistor pairs of the circuit. The circuit was then automatically placed with the algorithm described above and the resulting placement is shown in Fig. 4.25. The thermal profile of the placement is given in Fig. 4.26. Table 4.5 shows the performance characteristics of the operational amplifier. For each performance characteristic, the specification, nominal value and simulated value after placement is given. The nominal value of a characteristic is determined by simulating the circuit without parasitics. The value of a characteristic after placement is determined by simulating the circuit with the parasitic effects. Since the real values of interconnect parasitics are unknown, these values have to be be estimated. It can be concluded from Table 4.5 that all performance characteristics are within the specifications after placement. Table 4.6 explains the result for the offset voltage in more detail. This table shows the sensitivities, the temperature and the resulting offset voltage degradation for the most sensitive transistor pairs in the circuit. It can be seen from the layout that the sensitive transistor pairs have been placed symmetrically with respect to the output transistors. Note that the thermal profile calculation can also be used interactively, for example when a manually generated layout is given and the designer wants to explore the impact of thermal effects.

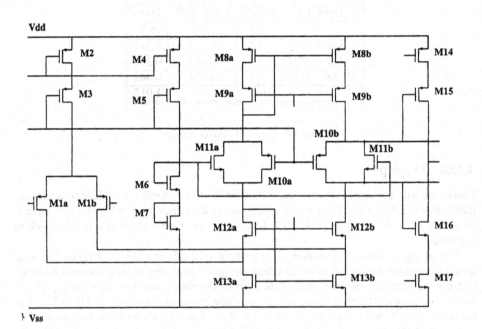

Figure 4.24: Opamp3 : schematic.

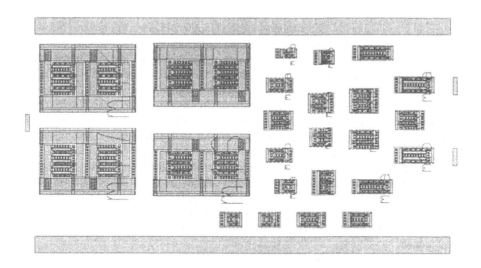

Figure 4.25: Opamp3 : placement

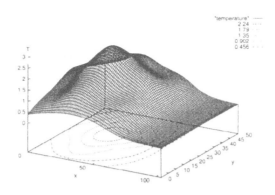

Figure 4.26: Opamp3 : thermal profile

4.13   Summary and Conclusions

Placement is a very important step in analog circuit layout. All layout parasitics that degrade the performance of a circuit, interconnect parasitics, device mismatch and thermal effects, are either determined or greatly influenced by the placement of the circuit. To keep the layout-induced performance degradation within user-defined margins, an automatic placement algorithm has to take into account all off these effects simultaneously. At the same time, a number of topological constraints, such as symmetry and aspect-ratio constraints have to be enforced. In this chapter, we have presented the first performance driven placement algorithm that takes all parasitic layout effects into account simultaneously and directly, without an intermediate constraint generation step.

The algorithm uses a simulated annealing algorithm to optimize the placement while keeping the layout-induced performance degradation within the specifications. The cost assigned to an intermediate placement is based on an accurate estimation of the performance degradation caused by the combined effect of interconnect parasitics, device mismatch and thermal effects. A new efficient and accurate thermal modeling technique has been developed to compute placement thermal profiles.

A number of industrial circuits have been placed with this algorithm. These experiments have shown that compact, fully functional placements can be generated in reasonable CPU times. Comparisons with placements produced by traditional, non-performance driven algorithms show that significant performance gains are obtained with only small increases in CPU time.

Chapter 5

Routing

5.1   Introduction

In this chapter, we discuss the routing problem for high-performance analog circuits. The routing phase is critical for the overall performance of the circuit, since it fixes the final values of the interconnect parasitics. While the placement phase has taken into account the effect on the performance of the minimum values for the interconnect parasitics, their real value is determined during routing. Therefore, the main concern during performance driven routing is to connect all wires while limiting the performance degradation introduced by the actual interconnect parasitics within the specifications of the user.

5.2   Problem Formulation

The analog performance driven routing problem can be stated as follows : given a placement of an electrical circuit specified as a set of devices together with their positions, orientations and implementations and a netlist interconnecting terminals on these devices and on the periphery of the placement itself (input/output pins, power supplies), find the layouts of all the nets such that the circuit is interconnected according to the netlist in a design rule correct way. For high performance analog circuits, the following additional constraints and objectives have to be added :

- **performance constraints**
  Each wire that is implemented by the router introduces a parasitic series resistance and a parasitic capacitance to every other conductor in the circuit. We will use an RC model for the interconnect and ignore inductive effects (see chapter 2). These elements introduce parasitic loading and coupling effects into the circuit. Routing has to be done such that the combined influence of these interconnect parasitics on the performance of the circuit remains within the specifications imposed by the designer.

- **symmetry constraints**
  For the differential signal paths in an analog circuit, the matching between parasitics on

symmetric nodes is often more important than their absolute values. To match the parasitics of two nets, they have to be routed symmetrically, even if the placement is not completely symmetrical.

- **variable wire width**
  Different parts of an analog circuit often carry significantly different currents. To avoid electro-migration in high-current wires, the widths of high-current wires must be increased.

- **yield/testability effects**
  The layout of a circuit has a profound impact on the likelihood of certain faults and fault types [Maly 90]. By careful routing, it is possible to reduce the total expected number of faults and hence to improve the yield of a circuit. Moreover, it is also possible to improve the testability of a circuit by decreasing the probability of occurrence of hard to detect faults. A manufacturability driven routing algorithm minimizes the defect sensitivity and increases the testability of an analog circuit, while bounding the performance degradation within the allowed margins.

- **n-layer routing**
  State-of-the-art technology processes have 6 or more metal layers. An analog routing algorithm has to take advantage of all available routing layers to improve the performance of the circuit.

5.3   Overview of the Routing Tool

The flow diagram of the analog routing tool is shown in Fig. 5.1. The input of the tool consists of a netlist file, placement information, a set of circuit specifications and a technology file. Before the actual routing tool is started, a circuit analysis tool is called (see chapter 2). This tool interfaces to a numerical circuit simulator and determines the performance sensitivities and operating point information for the circuit.

After circuit analysis, a pre-routing step is executed to estimate the impact of each net's parasitics on the performance of the circuit. Based on these estimations a routing order is determined (see section 5.8) and a performance-driven routing algorithm is used to wire the layout such that the performance degradation introduced by layout parasitics remains within the specifications imposed by the designer. A line expansion algorithm [Heyns 80] is used as the basic path finding mechanism.

If the performance-driven routing phase was successful and there is enough performance margin left, a yield and testability optimization loop is entered. During this loop, nets are removed and rerouted until the available performance margin is consumed or no further yield/testability improvement is found.

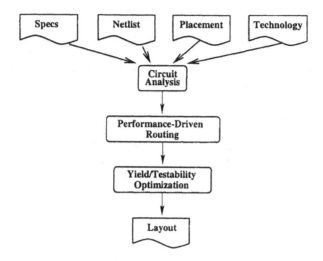

Figure 5.1: Overview of the analog routing tool.

5.4 Classification of Routing Algorithms

Routing has been an important problem in design automation for decades. Numerous routing algorithms have been presented in the past. To give an overview of these algorithms is beyond the scope of this book. In this section we will present some important concepts that can be used to classify routing algorithms and we will discuss their relation to the analog circuit level routing problem.

5.4.1 Routing Strategy

In routing of VLSI circuits, two strategies can be followed : *single-phase routing* and *two-phase routing*. The single-phase approach to the general routing problem is called *area routing*. In this approach, nets are routed one net at a time, considering the whole layout area. In the two-phase approach, the general routing problem is solved in two separate steps : *global routing* and *detailed* routing. In this approach, the space not occupied by circuit blocks is divided into *routing regions*. This includes spaces between blocks and above blocks, also called over the channel (OTC) areas. Between blocks there are two types of routing regions : channels and 2D-switch-boxes. Above blocks, the entire routing space is available and can be partitioned into smaller regions called 3D-switch-boxes. During global routing, a "loose" route is constructed for each net by assigning it to a list of routing regions. This list specifies a trajectory for each net, without defining the actual geometric layout of the wires. In the second phase, detailed routing, the actual geometric layout of the wires is generated by routing each region separately. Dedicated routing tools are used for each type of region : *channel routers*, *2D-switch-box routers* and *3D-switch-box routers*.

The two-phase approach has become a de facto standard for chip level routing problems. Nowadays, VLSI chips may contain several millions of transistors and tens of thousands of nets which have to be routed to complete the layout. This kind of complexity is beyond the capabilities of an area router and can only be handled by dividing the problem into smaller sub-problems which can be solved one at a time. For analog circuit level layout however, the problem size is usually much smaller. The number of interconnections that have to be routed in an analog routing problem is in most cases restricted to 10 to 20 nets, so the speed and efficiency of the routing algorithm is less important. Not only is the main advantage of the global routing/detailed routing approach not pertinent to the circuit-level analog routing problem, it also has a negative attribute which makes it even less attractive for analog use. By dividing the routing problem into smaller sub-problems, the global view of the problem is lost during detailed routing. This makes if difficult to impose critical analog constraints like symmetry, matching and crosstalk minimization. Therefore, we have selected an area routing algorithm for our layout tool.

5.4.2   Routing Model

In routing of VLSI circuits, several different approaches can be used to represent the routing problem. These approaches differ in the level of abstraction that is used, trading off efficiency and ease of implementation against generality and optimality of the solution. The way the routing problem is represented is referred to as the *routing model*. In this section, different routing models will be presented and compared, and the most appropriate model for analog routing will be selected.

A first classification of routing models is based on the level of abstraction that is used in the representation of the wires. In a *grid-based* model, a rectilinear grid is super-imposed on the layout area and the wires are restricted to follow paths along the grid lines. Wires are represented as paths without any thickness and the spacing between the grid-lines is used to accommodate for the actual wire thickness and spacing in the layout. In a *grid-less* routing model, the terminals, wires and vias are represented as polygons on different layers. The coordinates of the polygons are not restricted to a grid. Fig. 5.2 illustrates the difference between the two routing models.

Although the use of a grid makes computations easy, it also implies a number of restrictions which can not be tolerated in analog circuit level routing. First, since the spacing of the grid lines is determined by the worst-case spacing requirement over all layers, it leads to a considerable waste of area. Second, variable wire widths, which are an important requirement for analog routing, can not be used since the grid-based model requires a uniform pitch. Finally, the terminals are required to align on the grid lines, which is usually not the case for analog circuits. Therefore, a grid-less routing model has been selected for our routing program.

The second classification of routing models is based on the layer assignments of horizontal and vertical segments of nets. In an *unreserved layer model*, any net segment is allowed to be placed in any layer. *Reserved layer models* restrict certain types of segments to particular layers. An example of a reserved layer model is the VH model, where the first layer is reserved for vertical segments and the second layer for horizontal segments. The use of a reserved layer model is a restriction which is introduced to reduce the complexity of a problem (for instance to

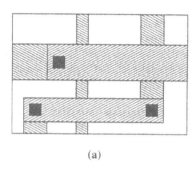 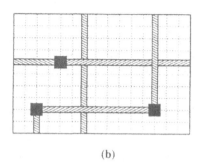

(a)   (b)

Figure 5.2: Different routing models :
    (a) grid-less
    (b) grid-based

avoid blocking situations) and to facilitate the computations. On the other hand, it also restricts the number of solutions that can be reached and may lead to sub-optimal solutions.

Generally speaking, reserved layer and gridded routing models reduce the complexity of the routing problem and will result in faster routers. However, to achieve their efficiency, they model the routing problem in a way that is too restrictive for analog circuit level layout. In addition, analog circuit level layout is usually of a much smaller complexity than general IC layout. Therefore, we have chosen a grid-less, unreserved layer model for our router.

5.4.3 Search Strategies

To connect a source terminal to a target terminal, an area router starts by generating a set of partial solutions from the source terminal and iteratively expands one partial solution into a number of new ones until the target is reached. This process can be modeled as a search problem in the graph of partial solutions. The order in which partial solutions are generated and evaluated is determined by the *search strategy*. Several search strategies can be used to explore a graph [Thul 92] :

- **Depth-First Search (DFS)** In a DFS strategy, a new solution is generated from the *most recently* generated solution. A partial solution is expanded as deeply as possible until the target is reached or no further expansion is possible. In the latter case, the algorithm tracks back to the previous partial solution and restarts the expansion process from there. The order of exploring solutions in DFS is last-in-first-out.

- **Breadth-First Search (BrFS)** In a BrFS strategy, all solutions on the same level are explored before any other solution is generated. If the target is not reached for any solution of expansion level n, the algorithm proceeds by generating solutions of level $n + 1$. The order of exploring solutions in BrFS is first-in-first-out.

- **Best-First Search (BeFS)** The basic idea of the BeFS strategy is to expand new solutions from the current solution with the best cost value. The advantage of this search strategy is that if the target is reached, we can be sure that the path found is the minimal cost path, since all other partial solutions visited have greater cost. BeFS can show a dramatic improvement in time and space efficiency over *blind* searches as DFS and BrFs.

- **Heuristic Search (HS)** BeFS relies on historical information to predict which partial solutions are the most likely to be on a minimal cost path. The ideal algorithm would operate on perfect information, thereby always choosing the correct solution to expand at each stage of the search. This is impossible, since the solution is not known. It is however possible to use heuristics to predict the solution which is most likely on a *minimal cost path*. In HS, this solution is selected for further expansion. An example of a heuristic search algorithm is the A^* algorithm [Nils 71].

In our routing tool we have chosen a heuristic search strategy with a special heuristic that targets the expansion towards the routing solution that is best for the circuit performance. This will be explained in detail in section 5.7.

5.5 Previous Work in Area Routing

In the previous section, it was shown that area routing is the most appropriate routing strategy for analog circuits. In this section we will discuss the most important algorithms that can be used for area routing, using the concepts explained in the previous section.

5.5.1 Maze Routing

Maze routing algorithms search for a path between a source and a target on a grid-based routing model. The layout area is divided into a rectangular grid of cells, with some cells free and others blocked. The search is started from the grid cells that are part of the source terminal. New solutions are generated by expanding the path outward to neighboring free cells. The search is terminated if a grid cell is reached that is part of the target terminal. Maze routers can be classified based on the search strategy that they use :

- **Lee's algorithm [Lee 61]** In this algorithm, the search is conducted symmetrically in every direction, using the breadth-first search technique. This can be seen as a wave propagating from the source until the target is reached. This algorithm guarantees finding the shortest path between two terminals if one exists.

- **Soukup's algorithm [Souk 78]** Soukup's algorithm uses a depth-first search until an obstacle is encountered. If an obstacle is encountered, a breadth-first search method is used to get around it. The running time of Soukup's algorithm is usually better than Lee's, but this algorithm does not guarantee to find the shortest path.

- **Hadlock's algorithm [Had 75]** Hadlock's algorithm uses the A^* heuristic search method to prefer the direction of the search toward the target. Hadlock's algorithm is usually faster than Lee's and Soukup's and guarantees finding a shortest path between the terminals if one exists.

Several extensions of these algorithms have been proposed to deal with multi-terminal nets and multi-layer routing models.

5.5.2 Line-Search Routing

Like maze routers, line-search routers search for a path between a source and a target terminal on a grid-based routing model. The essential difference between the two is the way in which a partial path is expanded into a new one. Whereas maze routers expand a partial path by adding a grid cell to it, line-search routers expand a path by adding a line segment to it. Since a line segment can potentially represent many grid cells, a considerable memory and speed advantage can be achieved. The basic operation of a line-search router is as follows. Initially, horizontal and vertical lines are drawn through the grid-points representing the source and the target terminal. These lines, called trial lines, are extended until they hit an obstacle or the external border of the routing plane. If a line originating from the source terminal intersects a line originating from the target terminal, the search is terminated and a path is constructed. Otherwise, new paths are generated from the previous ones by drawing lines perpendicular to the trial lines of the previous step. This process is repeated until a line from the sequence originating from the source terminal intersects a line from the sequence from the target terminal. Different search strategies again lead to different line-search routers :

- **Mikami/Tabuchi's algorithm [Mika 68]** In this algorithm, new trial lines are generated by drawing perpendicular line segments through every grid point of the current trial line. This search process is similar to breadth-first search and is guaranteed to find a path if one exists. However, the path may not be the shortest one.

- **Hightower's algorithm [High 69]** In Hightower's algorithm, only one new trial line is generated from a previous one. The new trial line is drawn such that it avoids the obstacle that blocked the current trial line. [High 69] describes three different procedures to avoid different types of obstacles. Although this search strategy performs very well in most practical cases, it can not guarantee that a path will be found when it exists.

5.5.3 Line-Expansion Routing

The line-expansion algorithm [Heyns 80] is a grid-based technique that combines elements from the maze routing and the line-search routing algorithms. The algorithm is based on expanding a line in its perpendicular direction into an *expansion zone*. The *expansion zone* is defined as the zone consisting of all grid points that can be reached by a line beginning on the expanded line and perpendicular to it. The boundary segments of the expansion zone, called *active lines*, are pushed onto a stack. In the next step, the generated active lines are expanded outside the zone for further search. This procedure is initiated from the source terminal as well as from the target terminal. A sequence of expansion zones is then generated from each of the terminals and the execution of the algorithm is terminated if a zone originating from the source terminals meets a zone originating from the target terminal. A back trace is then executed to find the connecting path. This algorithm guarantees to find a path if one exists. This may not be the minimum cost path.

5.5.4   Discussion

The area routing algorithms presented in this section are grid-based in their original formulations. As discussed in the previous section, the grid-based routing model is a disadvantage in the context of analog routing since it puts a severe and intolerable limitation on the possible solutions. However, the basic concepts of the maze routing, line-search and line-expansion algorithms can easily be generalized. The notion of expanding partial paths to neighbors is not restricted to rectangular grids but can be extended to any routing model. In a grid-less area router, partial paths are represented by their actual geometry and a general expansion process is defined to expand a partial path into its neighbors. If a grid-less routing model is used, the distinction between maze routing and line-search (or line-expansion) becomes less important. If a unit length expansion is used, the resulting routing algorithm can be called a *grid-less maze router*. If paths are expanded over longer distances or until they hit an obstacle, the term *grid-less line-search* router might be appropriate. The basic path finding mechanism used in our routing tool is a grid-less area routing algorithm that combines features of the different area routing algorithms presented above. The algorithm is similar to the ones presented in [Sato 87, Marg 87, Arno 88, Cohn 91]. Following [Marg 87] we will call our algorithm a grid-less maze routing algorithm, although it combines elements from the maze routing and line-search routing techniques. The algorithm will be discussed in detail in the next section.

5.6   A Grid-Less Maze Routing Algorithm

The basic operation of the algorithm that is used to connect a source and a target region is illustrated in Fig. 5.3. The router starts with a collection of partial paths that are derived from the source region (see section 5.6.2). These paths are sorted according to a cost function and stored in the collection of partially completed paths. The following procedure is then executed until one of the partial paths reaches the target region :

1. Out of the collection of partially completed paths, one path is selected for expansion. This path selection process is based on the A^* search strategy. The path selection process and the cost function will be discussed in section 5.7.

2. The partial path selected in step 1 is expanded into a collection of new partial paths using the expansion mechanism discussed in section 5.6.3.

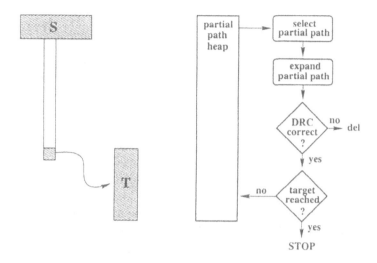

Figure 5.3: The line expansion algorithm.

3. The partial paths generated in step 2 are checked to see if they are design rule correct, i.e. if they do not overlap with the devices or with previously routed wires.

4. The design rule correct paths are checked again to see if they overlap with the target terminal. If they do, the connection is complete and the iterative procedure can be terminated. If the target has not been reached yet, the new partial paths are inserted into the partial path collection and they become candidates for further expansion, together with all the partial paths generated during previous iterations.

In the following sections, we will discuss the routing model and the path expansion steps in detail.

5.6.1 Routing Model

The routing model that is used in our routing algorithm is a grid-less model. The layout objects involved in the routing problem, devices, terminals and wires, are represented with their exact layout geometry.

5.6.1.1 Corner stitching

The layout area is represented as a collection of geometric shapes on different process layers. During each iteration of the routing algorithm, a design rule check and an overlap check has to be executed for each expanded path. Each of these operations involves an area search in the database of shapes representing the devices and previously routed wires. Since these database

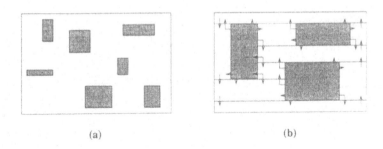

              (a)   (b)

Figure 5.4: Layout representations : (a) bin based (b) corner stitched.

access operations are executed thousands of times, an efficient data-structure, which allows fast area search, is needed. Several data-structures have been proposed used to store a collection of layout shapes :

- **linked list** The simplest data structure used to store a collection of shapes is a linked list, where each list element represents a shape. The complexity of the area search operation for a linked list data-structure is $O(n)$ where n is the number of shapes.

- **bin based** (see Fig. 5.4(a)) A bin based data structure can be seen as an augmented version of the linked list. A virtual grid is superimposed on the layout area. This grid divides the area into a series of bins which can be represented using a two-dimensional array. For each bin, a list of shapes intersecting it is stored. The worst case complexity for the area search operation is $O(b + n)$ where b is the number of bins. However, in practical situations, the effective complexity is a lot better than $O(b + n)$.

- **corner stitching data-structures** (tile planes, see Fig. 5.4(b)) [Oust 84] In corner stitching, shapes and empty space are represented by non-overlapping, rectangular *tiles*. Tiles are linked to their neighbors at their lower left and upper right corners by pointers, called *corner stitches*. Corner stitching differs from linked lists and bin based data-structures in that empty space is represented explicitly by space tiles. The worst case complexity of the area search operation is $O(n)$, but in practical situations, the area search operation is very fast, especially if so-called *hint* tiles are used.

Tile planes perform very well for area searches in practice, despite their theoretical worst-case running time that is proportional to the number of rectangles in the database. This makes them very attractive for use in grid-less routing algorithms and several area routers based on tile planes have been reported [Marg 87, Cohn 91].

    In our routing tool, we use a tile based area representation with two separate tile planes for each routing layer. Prior to routing, all features on a certain layer are inserted in two separate tile planes, one organized in maximal vertical strips and one organized in maximal horizontal strips. For each routing layer, the user can define additional areas where routing is not allowed.

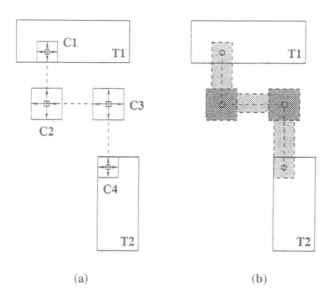

Figure 5.5: Representing paths by routing cells :
(a) routing cell representation
(b) physical path

These areas are also stored in the tile planes for the routing layer. Vias are duplicated in the tile planes of the two routing layers that they connect. The dual corner stitching representation for each layer speeds up the path expansion and the DRC steps during routing.

5.6.1.2 Routing Cells

During routing, partially completed wires are represented as chains of routing cells. Each routing cell is the result of an expansion and contains a back trace pointer to the routing cell from which it originated. This pointer allows to reconstruct a partial path from its head routing cell by following all back tracing pointers. A partial path can thus be represented by its head routing cell. Note that one routing cell can be part of many partial paths. Each routing cell has a cost associated with it. This cost is the sum of two terms : the total accumulated cost of the partial path that is represented by it and an estimate of the cost required to complete the path to the target terminal. There are two types of routing cells : regular cells and via cells.

A regular routing cell is the result of an expansion on the same layer and is represented by its (x, y) center-point, its layer number L and its width W. A via cell represents a connection between two consecutive routing levels and is represented by its (x, y) center-point, bottom and top layer numbers L_b and L_t and bottom and top widths W_b and W_t.

Fig. 5.5 shows an example of a path represented by routing cells. In this figure, the path

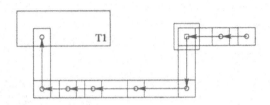

Figure 5.6: Reconstructing a partial path from its routing cell representation.

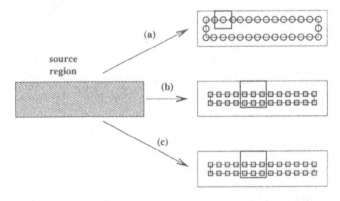

Figure 5.7: Source region expansion.
  (a) at the same layer
  (b) and (c) to a lower or higher layer through vias

connecting terminals $T1$ and $T2$ is represented by the four routing cells ($c1$, $c2$, $c3$, $c4$). Cells $c1$ and $c4$ are regular cells, $c2$ and $c3$ are via cells. In the remainder of this chapter, we will represent regular routing cells by circular symbols and via cells by square symbols on their center point. Fig. 5.6 shows how a partial path is reconstructed from its routing cell representation : starting from the head cell, we follow back trace pointers until the path changes direction or layer. The bounding box of all cells encountered during this process is added to the path layout. The last cell encountered is now used as a head cell and the process is repeated until the source cell is encountered. When a via cell is encountered, a physical via is constructed and added to the path layout. Note that routing cells are allowed to overlap each other.

5.6.2 Source Region Expansion

The basic path finding algorithm routes a wire between two connected layout regions, called the source and the target region. A region represents a connected subnet and consists of a set of layout features on different layers, connected through vias. A region can be a single device terminal or a partially connected subnet resulting from previous routing steps. In the latter case, the source region consists of a number of device terminals connected with a partially routed net. Before the path finding algorithm can start, this region has to be transformed into a collection of routing cells, suitable for further expansion by the algorithm. For each source box, source cells are generated on the same routing layer of the box, and on the next higher and next lower layer if the box dimensions allow the insertion of via cells. This process is depicted in Fig. 5.7.

Source cells on the same layer are created on the sides of the box that are adjacent to routable area. To determine the location of the routing cells, the source box is shrunk by half the cell width. Routing cell centers are placed on the perimeter of the shrunk box, at fixed intervals given by the minimum resolution of the technology process (see Fig. 5.7(a)). Note that all source cells overlap the box from which they originate.

If the dimensions of the box allow the insertion of a via, via cells connecting to the next higher and lower layers are also created and inserted. The centers of the via cells are again determined by shrinking the source box by half the via cell width and placing center points at fixed intervals given by the process resolution (see Fig 5.7(b) and (c)).

After source region expansion, the cost of each cell is computed using the cost function described in section 5.7 and the cells are inserted into the heap as candidates for further expansion.

The width of a routing cell depends on the current that is flowing in the wire that is to be routed. Let I be the current of the wire, then the width W_i of a regular routing cell on layer i is given by :

$$W_i = \frac{I}{I_{max,i}}, \tag{5.1}$$

where $I_{max,i}$ is the maximum current density of layer i, expressed in Ampere per meter. For a via cell connecting layers i and j, the computation has to be done based on the maximum current in a via. The neccesary number of vias N_{via} can be computed as follows :

$$N_{via} = \frac{I}{I_{max,ij}}, \tag{5.2}$$

where $I_{max,ij}$ is the maximum current in a via connecting layers i and j. The width W_{ij} of the via cell is then given by :

$$W_{ij} = \left[ceil(\sqrt{N_{via}}) - 1 \right] viaSeparation_{ij} + 2max(viaOverlap_i, viaOverlap_j), \tag{5.3}$$

where $ceil(x)$ denotes the smallest integer greater than or equal to x and $max(x, y)$ the maximum of x and y. $viaSeparation_{ij}$ and $viaOverlap_i$ are technology constants specifying the minimum separation between vias connecting i and j and the minimum layer i overlap of a via, respectively. Note that the wire width is a property of a wire segment, not of the entire net.

5.6.3 Path Expansion

During each iteration of the line-expansion algorithm, the best partial path is selected and expanded on the same and on different layers. Since a partial path is defined by its head routing

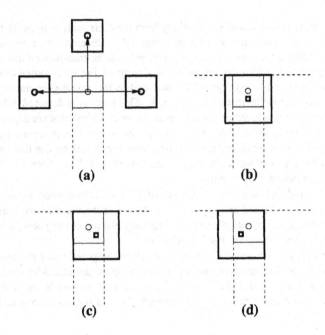

Figure 5.8: A regular routing cell expanded into new cells :
   (a) 3 regular routing cells on the same layer,
   (b) a via cell, aligned to the center of the original cell,
   (c) a via cell, aligned to the left edge of the original cell,
   (d) a via cell, aligned to the right edge of the original cell.

Figure 5.9: A via cell expanded into new cells :
   (a) regular routing cells on the top or bottom layer,
   (b) via cell to the next higher of lower level.

cell, this can be done by popping the cheapest routing cell from the heap and deriving a set of new routing cells from it. Regular routing cells are expanded into cells on the same layer and into one or more via cells. A via cell is expanded into regular cells on one of its layers and into one or more vias.

The expansion process for regular routing cells is illustrated in Fig. 5.8. Without loss of generality, we assume that the current routing cell was expanded from the bottom direction. Original cells in Fig. 5.8 and 5.9 are shown in thin lines, new cells in thick lines. For each expansion type, only one new cell is drawn, others are represented by their center points. New cells on the same layer are created by translating the original cell in all three non backward directions as shown in Fig. 5.8(a). Expansion on the next higher/lower layer is done by creating a number of via cells on top of the original cell. In general, the width of the via cells differs from the width of the regular routing cells and therefore, several different alignments are possible. The via cell is always created such that its top edge aligns with the top edge of the original cell. The three different side edge alignments are illustrated in Fig. 5.8(b) (center alignment), (c) (left edge alignment) and (d) (right edge alignment). In the most general case, the expansion of a regular routing cell results in 9 new cells : three regular cell on the same layer, three via cells on the next higher layer and three via cells on the next lower.

Via cells are expanded as shown in Fig. 5.9. We assume that the via cell from which the expansion is done originated from a lower layer, i.e. new regular routing cells are created on the top layer of the via cell. Since a via cell is in general larger than the corresponding top layer routing cell, expansion on the same layer is similar to source region expansion (see section 5.6.2). The layer box of the via is shrunk by half the routing cell width and top layer routing cell centers are placed on the perimeter of the shrunk box, at fixed intervals given by the minimum resolution of the technology process (see Fig. 5.9(a)). If the process allows stacked vias, via cells to the next higher layer can be created. If the size of the new via cell is smaller than the original one, the expansion is similar to the one used for regular routing cells (see Fig. 5.9(b)). In the opposite case, different alignments must be considered and the approach is the same as for creating new via cells from regular routing cells.

Each new routing cell results in a new partial path which is checked for design rule violations with respect to itself and to other already routed nets (DRC operation). If the new partial path is design rule correct, its cost is computed and the routing cell is inserted into the heap. The expand/DRC/compute cost sequence is executed thousands of times during the routing procedure. It is therefore crucial to perform these operations as fast as possible. The DRC operation involves an area search in the corner stitching database of the terminal and previously routed net geometry. The worst case complexity of this operation is $O(n)$, where n is the number of tiles in the database [Oust 84]. In practical situations however, the area search operation is very fast, especially if so-called *hint* tiles are used. The use of a hint tile reduces the complexity in most practical cases to $O(1)$.

Each new path expansion is done from the cheapest routing cell present in the collection of previously expanded cells. This is implemented by storing all routing cells in a priority queue and popping the cheapest cell from the queue for each expansion cycle. The priority queue is implemented by a Fibonacci heap [Fredman 87]. Inserting a cell in the heap takes time $O(logn)$, where n is the number of cells already present in the heap. The space requirement is $O(n)$.

5.7   Cost Function

In this section, we describe the cost function that is used to select the best partial path for further expansion. Consider a partial path P_n that is the result of n previous expansions. As described in section 5.6.1.2, P_n is represented by a chain of n routing cells c_i, where i denotes the ith expansion step :

$$P_n = [c_n, c_{n-1}, ..., c_2, c_1, c_0] . \tag{5.4}$$

c_{n-1} to c_0 can be derived from c_n by following back-trace pointers. c_0 is a source cell derived from the source region. The cost of P_n, $PathCost(P_n)$, is the sum of two terms :

$$PathCost(P_n) = PathCost_{act}(P_n) + PathCost_{pred}(P_n). \tag{5.5}$$

$PathCost_{act}(P_n)$ is the actual cost of the path from the source cell c_0 to the head cell c_n and $PathCost_{pred}(P_n)$, the predictor term, is an estimation of the cost required to complete the route from cell c_n to the target terminal. The predictor term is used to bias the expansion preferentially towards the target as described in section 5.7.2.

$PathCost_{act}(P_n)$ is computed recursively by adding the cost of the segment connecting c_n and c_{n-1} to $PathCost_{act}(P_{n-1})$:

$$PathCost_{act}(P_n) = Cost_{act}(c_n, c_{n-1}) + PathCost_{act}(P_{n-1}) \tag{5.6}$$

The actual cost of P_0, i.e. the path formed by the source cell c_0 is equal to zero :

$$PathCost_{act}(P_0) = 0 \tag{5.7}$$

The second term of the cost function, $PathCost_{pred}(c_n)$, can be computed from the location of the head cell of the path :

$$PathCost_{pred}(P_n) = Cost_{pred}(c_n) \tag{5.8}$$

Note that $PathCost_{pred}(P_n)$ is only a function of the head cell of the path, whereas $PathCost_{act}(P_n)$ is a function of the all routing cells in the path.

5.7.1   Actual Path Cost

This section describes the computation of the term $Cost_{act}(c_n, c_{n-1})$ of equation (5.6), i.e. the actual cost of the new wire segment introduced during expansion n. This computation is illustrated in Fig. 5.10. The addition of routing cell c_n to P_{n-1} introduces an additional wire segment (c_n, c_{n-1}) on layer L_n, with area A_n, length l_n and width w_n. We assume that the connection we are routing is part of net i. The parasitics associated with net i are the parasitic resistance R_i, the parasitic capacitance to ground c_i and parasitic coupling capacitances c_{ij} to all other nets $j, j = 0, .., i-1, i+1, ..., N$. The contribution of segment (c_n, c_{n-1}) to each of these parasitics is computed next.

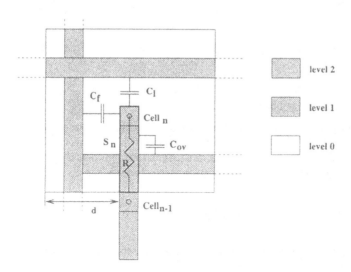

Figure 5.10: Parasitics introduced by the segment $c_{n-1} \rightarrow c_n$: parasitic resistance R, overlap capacitance C_ov, lateral capacitance C_l and fringing capacitance C_f.

The parasitic resistance R_i introduced by (c_n, c_{n-1}) is given by:

$$R_i = \rho_{square,L_n} \frac{l_n}{w_n} \qquad (5.9)$$

where ρ_{square,L_n} is the sheet resistance of layer L_n.

To compute the parasitic capacitances, a region of influence is defined around the segment as shown in Fig. 5.10. The extension distance d is a trade off between accuracy and speed. A large d implies a large area that has to be analyzed and hence requires greater computation times. A small d results in ignoring fringe capacitance contributions that might be important.

Within the region defined by d, three types of parasitic capacitance must be considered : overlap capacitance C_{ov} due to overlap between two segments on different layers, lateral capacitance C_l between two segments on the same layer and fringe capacitance C_f between segments on different layers. Each of these parasitic capacitance components can be computed from the geometric configuration of the layout. In our routing tool, we use the models reported in [Arora 96].

The lateral and fringing capacitances contribute to the coupling capacitance c_{ij} between the two nets involved. The overlap capacitance contributes to the capacitance to ground c_i or to the coupling capacitors c_{ij}, depending on the type of overlap.

The performance degradation ΔP_j for the performance characteristic P_j due to parasitics introduced by segment (c_n, c_{n-1}) can be determined using the pre-calculated sensitivity information:

$$\Delta P_j = S_{C_i}^{P_j} C_i + S_{R_i}^{P_j} R_i + \sum_{k=1, k \neq i}^{m} \frac{1}{2} S_{C_{ki}}^{P_j} C_{ki} \tag{5.10}$$

where m is the number of nodes in the circuit. $S_{C_i}^{P_j} = \dfrac{\delta P_j}{\delta C_i}$, $S_{R_i}^{P_j} = \dfrac{\delta P_j}{\delta R_i}$ and $S_{C_{ki}}^{P_j} = \dfrac{\delta P_j}{\delta C_{ki}}$ are the sensitivities of performance characteristic P_j to small changes in the parasitic capacitance C_i, the parasitic resistance R_i and the coupling capacitance C_{ki}, respectively. These sensitivities are determined in advance by simulation. The total cost of the segment is computed by summing the performance degradations ΔP_j for each performance characteristic P_j :

$$Cost_{act}(c_n, c_{n-1}) = \sum_{j=1}^{N_s} \Delta P_j \tag{5.11}$$

where N_s is the number of performance specifications for the circuit. This cost value can be combined with the contributions of all the previous segments to determine the total actual cost of the partial path P_n using equation (5.6).

5.7.2 Predictor Term

The purpose of the predictor term $PathCost_{pred}(P_n)$ in the cost function is to incorporate an estimation of the cost-to-completion in the cost of a partial path P_n. This ensures that, if a number of partial paths have equal actual cost values, the one which is most likely to be on a minimum cost path will be selected for expansion. It can be shown that this search strategy yields a minimum cost path if and only if the predictor term $PathCost_{pred}(P_n)$ is a lower bound for the actual value of a minimum cost path from the head cell of P_n to the target.

The computation of $PathCost_{pred}(P_n)$ is formally equal to the computation of the actual cost (5.10,5.11). The only difference is the way the parasitics are estimated. During the computation of the actual path cost, the exact geometry of the partial path is known and the parasitics can be extracted from the layout geometry. For the predictor term, the layout geometry is unknown and the wire is assumed to take on a minimum Manhattan path from the cell to the target terminal. Once this assumption is made, the situation is analogous to the estimation of interconnect induced performance degradation during placement optimization. We use the same technique to compute the estimated performance degradation and the reader is referred to section 4.9.1 for a detailed description.

5.7.3 Symmetric Routing

In analog circuits, it is often critical to match parasitics on symmetric nodes. To achieve this matching, nets have to be routed symmetrically, even if the placement is not completely symmetric. In our routing tool, symmetric routing is implemented using a strategy similar to the ones presented in [Malavasi 90, Cohn 91].

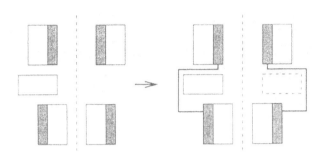

Figure 5.11: Symmetric routing.

A pair of symmetric nets is routed in one step. During this routing step, only one of the nets is actually routed, the symmetric counterpart of the net is generated afterwards by mirroring the net with respect to the symmetry axis. To make sure that the layout of the symmetric net pair is DRC correct on both sides of the axis, the DRC step which is executed after each expansion step has to be done on both sides of the axis. A new cell is first checked for DRC violations on the side that is actually routed. If it passes this check, it is mirrored to the other side of the symmetry axis and checked again. The cell is accepted as a candidate for further expansion only if it is legal on both sides of the symmetry axis. This double check procedure guarantees symmetric routes even in the case of partially symmetric placement.

The procedure is illustrated in Fig. 5.11. During routing of this symmetric net pair, only the net on the right side is actually routed. In the absence of symmetry constraints, this would result in an almost straight route between the two terminals. Due to the symmetric expansions process, the route has to make a detour to avoid the obstacle that is present on the left side of the placement.

5.7.4 Multi-Terminal Nets

The algorithm described in section 5.6.3 can be used to connect a source terminal to a target terminal. In general, a net has to connect an arbitrary number of terminals in an optimal way. Finding an optimal connection for a multi-terminal net is know as the Steiner tree problem and has been proven to be NP-hard. Several heuristic techniques can be used to find a sub-optimal solution. These algorithms apply a path connection algorithm, such as the one we presented in section 5.6.3 in an iterative way. Depending on the order in which terminals are connected, they can be classified in two categories.

The first group of multi-terminal connection algorithms is based on the idea of Prim's minimum spanning tree algorithm [Prim 57].

1. Randomly select a terminal from the set of terminals to be interconnected and use it as the source.

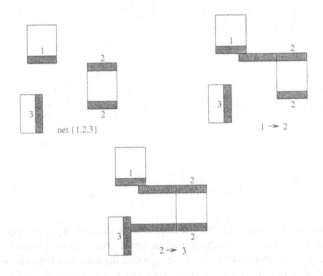

Figure 5.12: Routing multi-terminal nets.

2. Select the terminal which is closest to the source in Manhattan distance and use it as the target. Use the basic path connection algorithm to connect source and target terminal.

3. Create a new source by combining the source and target terminals and the routed connection and go to step 2.

The second algorithm is similar to Kruskal's spanning tree construction process [Krus 56] and consists of the following steps :

1. Find the two terminals which are closest to each other and connect them using the basic routing algorithm.

2. Unify the two terminals and their interconnection into one terminal and go to step 1.

We have implemented both of these algorithms in our routing tool. The quality of the nets produced by both algorithms was found to be similar in all our experiments. The routing sequence for a three terminal net is illustrated in Fig. 5.12. Note that in this case, both algorithms would use the same routing sequence.

5.8   Net Scheduling

In the previous sections, we described the algorithm that is used to find an optimal route for a net, i.e. a route that minimizes the performance degradation for the circuit. To route a circuit, this algorithm can be called in an iterative fashion until all nets have been routed. However, the sequential nature of this approach makes it very hard to predict the consequences that routing a net will have on the global performance and routability of the circuit. As a consequence, the quality of the final result depends heavily on the order in which the nets are routed. To reduce the dependence of the router on net ordering, and to improve the quality of the final result, we have introduced a three phase routing schedule consisting of a pre-routing step, a performance driven routing phase and a manufacturability improvement phase.

5.8.1   Pre-Routing Phase

The first phase is the pre-routing step, during which 'loose' routes for each net are determined. During pre-routing, each net is routed without taking into account design rule errors or constraint violations, i.e. we find the optimal route for each net, independent of all the other nets. Since most of the execution time of the routing algorithm is taken up by design rule checks and performance calculations, nets can be pre-routed very fast. While this preprocessing step introduces only a small computational overhead, it has several advantages for the performance driven routing step.

5.8.2   Performance Driven Routing Phase

After the pre-routing step, a performance optimization loop is entered. During this loop, nets are removed and rerouted until the resulting performance degradation is within the specifications imposed by the user. During rerouting, we use the cost function described in section 5.7 to determine the path of the wires. The order in which nets are ripped up and rerouted is determined by the results of the pre-routing step.

Based on the pre-routed circuit, we determine all parasitics $p_i(i = 1, ..., N_p)$. The total degradation ΔP_j of performance characteristic P_j can then be computed as:

$$\Delta P_j = \sum_{i=1}^{N_p} S_i^j p_i \tag{5.12}$$

The contribution ΔP_j^k of net k in ΔP_j is:

$$\Delta P_j^k = \sum_{i=1}^{N_p^k} S_i^j p_i \tag{5.13}$$

where N_p^k is the number of parasitics associated with net k. We now define F_k, the performance impact factor for net k as:

$$F_k = \frac{\sum_{j=1}^{N_s} \frac{\Delta P_j^k}{\Delta P_j}}{N_s} \tag{5.14}$$

where N_s is the total number of specifications imposed on the circuit. F_k is a number between 0 and 1 and measures the impact of parasitics associated with net k on the overall performance of the circuit. We use this factor to determine the net routing schedule. Nets are ripped up and rerouted in increasing order of performance impact factor, i.e. the nets with the highest impact on the performance are left in their pre-routed, optimal state, and nets with lower impact on performance are forced to take on less optimal paths to remove DRC violations and to improve performance.

5.8.3 Manufacturability Phase

If the performance-driven routing phase was successful and there is enough performance margin left, a yield and testability optimization loop is entered. During this loop, nets are removed and rerouted until the available performance margin is consumed or no further yield/testability improvement is found. The algorithm used for the rerouting of nets differs in two ways from the one used during performance-driven routing.

- Instead of minimizing performance degradation, partial paths that introduce specification violations are detected during the search and pruned from the search heap. In this way, it is guaranteed by construction that no specifications are violated during yield/testability optimization.

- The cost function is changed to favor paths that optimize the manufacturability of the circuit. A term that measures the impact on the manufacturability of the circuit is added to the cost function:

$$T_{manuf} = \sum_{i=1, i \neq k}^{n} \lambda_{ki} \Psi_{ki} \qquad (5.15)$$

with n the number of nets and k the net which is rerouted. In this equation, λ_{ki} denotes the expected number of bridging faults between circuit node k and i, and Ψ_{ki} is a measure of the difficulty of detecting the bridging fault. The derivation of λ_{ki} and Ψ_{ki} is discussed in the next section.

5.9 Estimating Yield and Testability

In the yield estimation literature, a defect is defined as anything that may cause a functional failure of a circuit. Various defect sources have been identified and analyzed [Maly 86] : wafer defects, human errors, equipment failure, environmental impact and process instabilities. Defects can be classified in two categories : global defects and local defects. Global defects are characterized by their overall influence on a complete wafer. Although some of the global defects will cause functional failure of the devices, most of them result in parametric yield loss. Local defects affect only a small localized region of a circuit and *may* result in functional failure of a circuit. Local defects that result in functional failures are called faults and their number is

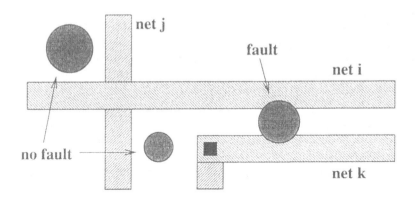

Figure 5.13: Layout fragment illustrating the influence of layout on defect-to-fault mapping.

determined by the layout of a circuit. In Fig. 5.13 a fragment of a layout with three local defects is shown. One of the three defects causes a short between net i and net k, the other two have no influence on the connectivity of the circuit. For this layout, three defects result in one fault. By increasing the separation between the wires implementing net i and net k, this fault can be avoided. In [Stapper 83a, Stapper 84] the concept of *critical area* was defined as the area of a layout in which the center of a defect must fall to cause a fault. In general, the critical area of a layout is a function of the defect size. The larger the radius of a defect, the larger the area in which the defect can lie to cause a fault. The critical area can be defined separately for each defect mechanism and for each fault.

5.9.1   Yield Modeling

If there are m possible defect mechanisms for a given process, the average number of faults $\lambda_{i,j}$ (sometimes referred to as *failure rate*) for a fault i due to a defect mechanism j can be obtained as :

$$\lambda_{i,j} = \int_0^\infty A_{i,j}(\chi) D_j(\chi) d\chi \tag{5.16}$$

where A_{ij} denotes the critical area for fault i due to defect mechanism j and $D_j(\chi)$ the defect size distribution for defect mechanism j. The average number of faults for fault i due to the combined effect of all defect mechanisms is then given by :

$$\lambda_i = \sum_{j=1}^m \lambda_{i,j} \tag{5.17}$$

Under the assumption of Poisson distribution of defects, the elementary sub-yield Y_i referred to fault i is [Stapper 83b] :

$$Y_i = e^{-\lambda_i} \tag{5.18}$$

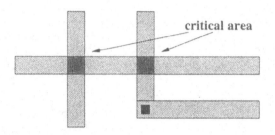

Figure 5.14: Critical area for dielectric pinhole defects.

Y_i is the probability of non-occurrence of fault i. The yield Y of the circuit is the probability that no fault is present in the circuit and can be computed as :

$$Y = \prod_{i=1}^{n} Y_i = \prod_{i=1}^{n} e^{-\lambda_i} = e^{-\lambda} \tag{5.19}$$

where $\lambda = \sum_{i=1}^{n} \lambda_i$. The probability P_i that a fault i occurs can be derived from (5.18) :

$$P_i = 1 - Y_i = 1 - e^{-\lambda_i} \tag{5.20}$$

Using equations (5.17) - (5.20), the problem of determining the probability of occurrence for each individual fault P_i and the overall yield Y for a given layout can be reduced to determining the defect size distributions $D_j(\chi)$, $j = 1 \cdots m$ for each defect mechanism and the critical area A_{ij} for each fault i and defect mechanism j.

We will consider two types of defect mechanisms in this work [Stapper 84]. First are *dielectric pinholes*, very small defects which often occur in insulators, like silicon dioxide or silicon nitride which are used between the conductive layers of integrated circuits. Their occurrence can result in a short between wires at different routing levels. The critical area associated with these defects is the overlap region between two wires (Fig. 5.14). If fault i is a short between two nets j and k, the critical area for fault i caused by dielectric pinholes $A_{i,ph}$ can thus be determined as the total overlap area $A_{ov,jk}$ between wires at different photolithographic levels implementing net j and net k :

$$A_{i,ph} = A_{ov,jk} \tag{5.21}$$

The size of dielectric pinhole defects is very small compared to layout dimensions, and can be considered constant for yield calculations. The defect size distribution is then simply given by :

$$D_{ph}(\chi) = \overline{D_{ph}} \tag{5.22}$$

where $\overline{D_{ph}}$ is the average density of dielectric pinholes, in units of defects per unit area. The average number of faults for fault i due to the dielectric pinhole defect mechanism can be calculated using (5.16) with the critical area given by (5.21) and the defect size distribution by (5.22) :

$$\lambda_{i,ph} = A_{ov,jk}\overline{D_{ph}} \tag{5.23}$$

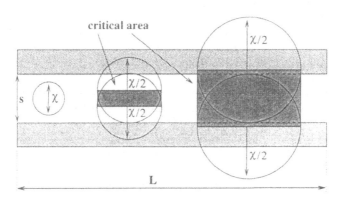

Figure 5.15: Critical area for a short between two wires caused by photolithographic defects.

The second class of defects are photolithographic defects, for which the defect size becomes of importance. Sub-micron pattern dimensions are typical for integrated circuits manufactured today. Dust and dirt particles with similar dimensions can interfere with the photolithographic processes used to define the patterns. These defects can break wires (*open* faults) and can create bridges between wires on the same routing level (*short* faults). The critical area calculation for a bridging fault i between two parallel conductors of length L, separated by a narrow slit of width s is illustrated in Fig. 5.15. It can be seen that the critical area is a function of the defect size χ. If χ is smaller than the separation s between the wires, the defect can not cause a bridging fault and the critical area is zero. If χ is larger than s, the critical area is proportional to χ. The critical area $A_{i,pd}$ for bridging fault i caused by photolithographic defects can thus be modeled by the following equation :

$$A_{i,pd}(\chi) = \begin{cases} 0 & \text{for } 0 \leq \chi \leq s \\ L(\chi - s) & \text{for } s \leq \chi \leq \infty \end{cases} \tag{5.24}$$

where L is the parallel length of the two wires (see Fig. 5.15). The critical area calculation for an open fault in a wire (see Fig. 5.16) is analogous to the calculation for a short between two wires. If w is the width of the wire, the critical area is given by :

$$A_{i,pd}(\chi) = \begin{cases} 0 & \text{for } 0 \leq \chi \leq w \\ L(\chi - w) & \text{for } w \leq \chi \leq \infty \end{cases} \tag{5.25}$$

The critical area as a function of defect size has to be combined with a defect density distribution to compute the average number of faults. An acceptable distribution was given in [Stapper 83a]:

$$D_{pd}(\chi) = \begin{cases} \frac{\chi \overline{D_{pd}}}{\frac{\chi_0^2}{2}} & \text{for } 0 \leq \chi \leq \chi_0 \\ \frac{\chi_0^2 \overline{D_{pd}}}{\chi^3} & \text{for } \chi_0 \leq \chi \leq \infty \end{cases} \tag{5.26}$$

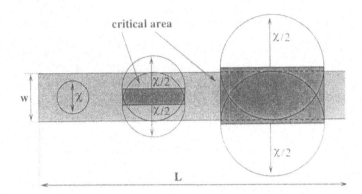

Figure 5.16: Critical area for an open in a wire caused by photolithographic defects.

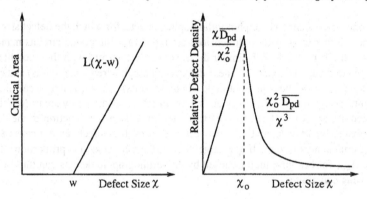

Figure 5.17: Critical area and defect density as a function of defect size

with $\overline{D_{pd}}$ the average photolithographic defect density in units of defects per unit area. This defect size distribution and the critical area are plotted in Fig. 5.17. The density of very small defects is assumed to increase linearly with defect size to a point where the straight line crosses the $\frac{1}{\chi^3}$ curve. The peak of this distribution occurs at defect size χ_0, which depends on the technology. Defects smaller than χ_0 can not be resolved by the optics used in the photolithographic process. The minimum dimensions of the patterns must therefore always be larger than χ_0. The average number of bridging faults $\lambda_{i,pd}$ due to photolithographic defects can be calculated by inserting the critical area (5.24) and the defect size distribution (5.26) in (5.16) :

$$\lambda_{i,j} = \int_s^\infty L(\chi - s)\frac{\chi_0^2\overline{D_{pd}}}{\chi 3}d\chi \tag{5.27}$$

The second part of the defect density distribution has been used since χ_0 is smaller than s. Eval-

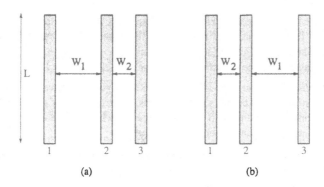

Figure 5.18: Layout fragments, with equal expected number of faults.

uation of (5.27) gives :

$$\lambda_{i,j} = \frac{L \chi_0^2 \overline{D_{pd}}}{2s} \qquad (5.28)$$

A similar result can be derived for opens :

$$\lambda_{i,j} = \frac{L \chi_0^2 \overline{D_{pd}}}{2w} \qquad (5.29)$$

where w is the width of a wire. The total failure rate λ_s for a fault s causing a short between two nets j and k can be computed by summing the failure rates caused by dielectric pinholes (5.23) and by photolithographic defects (5.28) :

$$\lambda_s = A_{ov,jk} \overline{D_{ph}} + \frac{L_{jk} \chi_0^2 \overline{D_{pd}}}{2s} \qquad (5.30)$$

where $A_{ov,jk}$ is the total overlap area between the two wires and L_{jk} the total parallel length of the two wires. The total failure λ_o rate for a fault o causing an open in a net j is given by :

$$\lambda_o = \frac{L_j \chi_0^2 \overline{D_{pd}}}{2w} \qquad (5.31)$$

with L_j the total length of the net.

5.9.2 Testability

Based on the expected number of faults, it is impossible to distinguish between the two layout fragments depicted in Fig. 5.18. In both cases, the expected number of faults is equal to $L \chi_0^2 \overline{D_{pd}} (\frac{1}{2W_1} + \frac{1}{2W_2})$. However, from a testability point of view, it is possible that faults between nodes 1 and 2 are easier to detect than faults between nodes 2 and 3 and in that case (b) is to be preferred over (a). To enable the router to make intelligent decisions in cases like this, we need a measure of detectability to weigh the faults. Such a measure is developed next.

Analog test strategies can be classified in two categories [Milor 89]:

- (a) *specification based testing* : this technique distinguishes between a good and a faulty circuit by testing all of a circuit's specifications. Specifications of analog circuits are typically based on their dynamic and transient behavior which makes testing for all specifications a time-consuming and expensive technique.

- (b) *fault based testing* : this technique detects faulty circuits by measuring a set of electrical responses to an input stimulus and comparing this set to the simulated responses of the fault-free circuit. This technique is much cheaper but has the disadvantage that circuits can be mis-classified. Due to the statistical nature of this test technique, faulty circuits can be classified as good ones and vice versa.

In the remainder of this section we will discuss only parametric testing techniques [Gielen 94]. We consider an electrical circuit whose response to an input stimulus is measured. The response measurements may be nodal voltages, currents, linear matrix parameters, etc. If m electrical responses are measured, the m-dimensional vector $\phi = (\phi_0, \phi_1, ..., \phi_{m-1})^T$ is called the response vector of the circuit. A typical parametric test strategy consists of the following steps :

1. During design, N Monte-Carlo simulations are carried out to measure the mean vector μ_0 and the covariance matrix Σ_0 of ϕ for the fault-free circuit. N can be chosen such that the standard deviations of the response estimates fall within a prescribed tolerance. During the Monte-Carlo simulations, normal distributions are assigned to all parameters affecting the performance of the circuit (technology parameters, device parameters, etc.)

2. During testing, the response vector ϕ is measured for the circuit under test and compared to the simulated response. Since a direct comparison is impossible, a statistical decision criterion must be used. Assuming that ϕ for the nominal circuit takes on a multivariate normal distribution, it can be shown that the solid ellipsoid of ϕ vectors satisfying the relation :

$$(\phi - \mu_0)^T \Sigma_0^{-1}(\phi - \mu_0) \geq \chi_m^2(\alpha) \qquad (5.32)$$

has probability α [And 58]. $\chi_m^2(\alpha)$ is the $100\alpha th$ percentile of a chi-square distribution with m degrees of freedom. To determine if the circuit under test is fault-free, the following null-hypothesis has to be tested :

$$H0 : E\{\phi\} = \mu_0 \qquad (5.33)$$

This test is performed by evaluating the left-hand side of (5.32) for the measured response vector ϕ. If the result is greater than $\chi_m^2(\alpha)$, the circuit is rejected and is faulty with probability $1 - \alpha$. If the result is less than $\chi_m^2(\alpha)$ the circuit is accepted and is fault-free with probability $1 - \alpha$. The probability of falsely rejecting a fault-free circuit is α. The value of α can be chosen depending on the application and the cost of rejecting good circuits [Gielen 94].

The effectiveness of the statistical decision criterion (5.32) depends strongly on the separability of the response vector distributions of the fault-free and the faulty circuits. The method yields accurate results when the response vectors of the faulty circuits lie in a region of response space which is separable from the circuit's fault-free response.

In a circuit with n nodes, there are in general $\frac{1}{2}n(n-1)$ types of shorts possible between the nodes. Denote by $\mu_{(k)}$ the mean of the response vector ϕ for the kth type of faulty circuit. The statistical distance between the response vector of the good circuit and that of the circuit with fault k can be defined as [And 58]:

$$d_{0,k} = (\mu_{(0)} - \mu_{(k)})^T \Sigma_{(0)}^{-1} (\mu_{(0)} - \mu_{(k)}) \tag{5.34}$$

The smaller the value of $d_{0,k}$, the higher the chance that fault k will remain undetected during testing. We can thus put Ψ_{ki} in equation (5.15) equal to the inverse of the statistical distance between the response vector of the good circuit and that of the circuit with nodes k and i shorted.

The cost function term (5.15) can thus be interpreted as the sum of the expected number of faults between net k and all other nets i, multiplied by the probability that they will remain undetected during testing. Minimizing this term for each net means optimizing the yield and the testability of the overall circuit.

The testability criterion developed above can be used with any analog parametric test method. To test the routing algorithm, we have used the test technique described in [Gielen 94]. In this technique, the results of time-domain simulations of the power-supply current are used to construct the signature of a circuit. The circuit response vector is equal to $\phi = (\phi_0, \phi_1, ..., \phi_{m-1})^T$, where ϕ_0 is the RMS value and $\phi_1, ..., \phi_{m-1}$ are the first $m-1$ harmonics of the power-supply current. The experimental results presented in the next section were generated with this testing algorithm.

5.10 Experimental Results

The above router has been implemented in C^{++} and is part of the performance-driven analog cell layout generation tool LAYLA, which itself can be used stand-alone or integrated with the analog synthesis environment AMGIE [Gielen 95a]. Two CMOS operational amplifiers are presented to demonstrate the effectiveness of the algorithm.

5.10.1 Opamp1

Opamp1 (see Fig. 5.19) is a moderate performance circuit and is therefore a good candidate for yield/testability optimization. During circuit analysis, Monte-Carlo analysis was used to determine the statistical distance between the I_{PS} spectrum of the good circuit and that of all the possible faulty circuits. The circuit was then automatically placed with the placement tool described in chapter 4 and then routed using the algorithm described in this chapter. The resulting layout is shown in Fig. 5.20. The performance and yield/testability characteristics for the circuit after layout are given in Table 5.1. The third column gives the results after the performance-driven stage of the algorithm and the fourth column gives the results after yield/testability optimization.

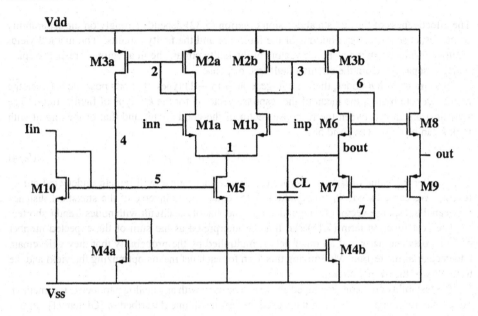

Figure 5.19: Opamp1 : schematic.

The performance characteristics given are the unity-gain bandwidth (UGBW) and the phase margin (PM). To evaluate the testability of the layout, a large number of circuits was tested using the I_{PS} monitoring technique. Every type of fault was represented in this test set with a probability calculated from the layout using the technique described in section 5.9. The Test Error Rate (TER) is defined as the percentage of faulty circuits that are classified as good circuits by the test algorithm. For this moderate performance circuit, the TER was driven to almost zero during the yield/testability stage, while the performance characteristics remained within the specifications.

Performance	Spec	Stage 1	Stage 2
$UGBW(MHz)$	> 1	1.013	1.010
$PM(^o)$	> 60	61.04	60.09
$Test\ Error\ Rate(\%)$		0.11	0.01

Table 5.1: Performance and Test Error Rate for Opamp1 after performance-driven routing (Stage 1) and after yield and testability optimization (Stage 2).

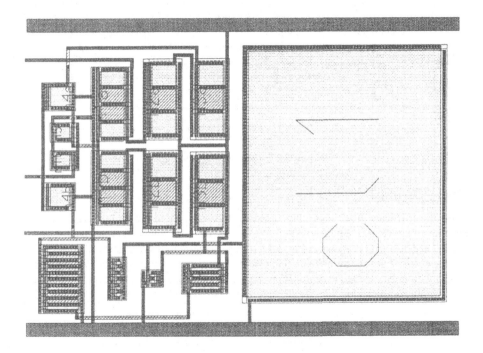

Figure 5.20: Opamp1 : layout.

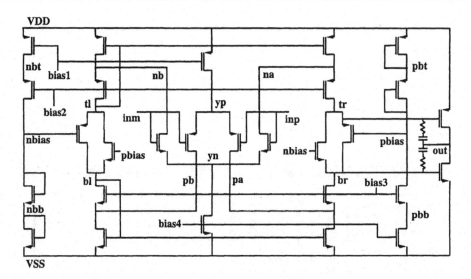

Figure 5.21: Opamp2 : schematic.

5.10.2   Opamp2

To test the efficiency of the algorithm, a second test was carried out with a larger and higher performance circuit :*Opamp2* (see Fig. 5.21). The layout was generated in the same way as for *Opamp1* and the result is shown in Fig. 5.22. The performance characteristics and the Test Error Rate for *Opamp2* are shown in Table 5.2. For this circuit, the Test Error Rate was significantly reduced with only a moderate degradation in performance. The performance of the final layout is still within specifications.

Performance	Spec	Stage 1	Stage 2
$UGBW(MHz)$	> 7	7.13	7.06
$PM(°)$	> 70	70.3	70.2
$Test\ Error\ Rate(\%)$		0.21	0.03

Table 5.2: Performance and Test Error Rate for Opamp2 *after performance-driven routing (Stage 1) and after yield and testability optimization (Stage 2).*

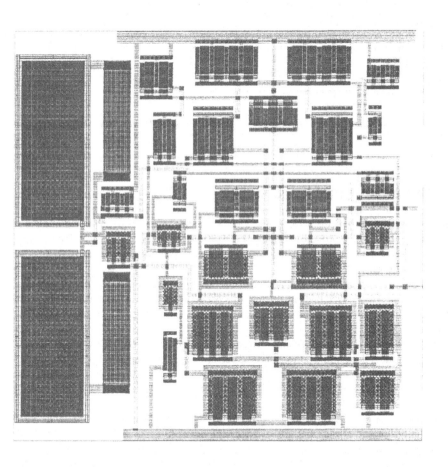

Figure 5.22: Opamp2 : layout.

5.10.3   CPU Times

The layouts for both test circuits were generated on an HP 712 workstation. The execution times for the individual steps and the total execution times are given in Table 5.3. It can be seen from this Table that the CPU-time for the additional yield/testability optimization step is acceptable for both circuits.

	Opamp1		*Opamp2*	
Program Step	seconds	% of total	seconds	% of total
performance-driven placement	184	33	263	33
performance-driven routing	210	38	321	40
yield/testability optimization	163	29	213	27
total CPU time	557	100	797	100

Table 5.3: Execution times for the different layout generation steps for test circuits Opamp1 *and* Opamp2

5.11   Summary and Conclusions

Routing has an impact on both the performance and the manufacturability of an analog integrated circuit. In this chapter, we presented an analog routing tool that maximizes the manufacturability of a circuit while keeping the performance within the user's specifications. To achieve this, we use a three step algorithm.

In a pre-routing step, we determine an optimal route for each net, independent of the other nets in the circuits. We analyze the parasitics of·this pre-routed circuit to estimate the impact of each individual net on the overall performance of the circuit and to determine a sensible routing schedule.

During the second phase, a performance driven routing algorithm is used to rip-up and reroute nets until the layout is design rule correct and the performance degradation is within the specifications imposed by the user. The cost function that drives this performance driven routing phase uses performance sensitivities to determine the impact of routing parasitics on the performance of a circuit.

In the third routing phase, any remaining performance margin is used to optimize the yield and the testability of the layout. Nets are rerouted with a cost function that is designed to favor paths that optimize the manufacturability of the circuit. A term that measures the impact on the manufacturability of the circuit is added to the cost function. This term is the product of the bridging fault probabilities between the routed net and all other nets, multiplied by a measure of their detectability.

Using two analog circuits, it was demonstrated that the use of our routing algorithm significantly improves the yield and the testability of the circuit layout while the performance remains within the specifications.

Chapter 6

Implementation

6.1 Introduction

This chapter briefly describes some implementation details of the LAYLA system, and some of our experiences with its introduction in an industrial environment.

6.2 Implementation

6.2.1 Source Code

The algorithms described in this book have been implemented in the C^{++} language in the UNIX environment. The total system comprises about 115000 lines of source code. The total system is organized in the following 8 subsystems :

- *basic data structures (6000 LOC)* : implementation of data structures that are used throughout the system : linked lists, dictionaries, heaps, etc..

- *parser (11000 LOC)* : code that parses the netlist file, the technology file and the performance specification files. This code was developed using the LEX and YACC parser generator programs.

- *technology (6000 LOC)* : data structures to store technology information and code to access and manipulate technology data.

- *geometry (5000 LOC)* : data structures and routines to store and manipulate geometric data. This includes the minimum spanning tree computation routine and the tile plane data structure used in the router.

- *circuit analysis (4000 LOC)* : code that interfaces to a commercial circuit simulator and processes the output to determine performance sensitivities and operating point information.

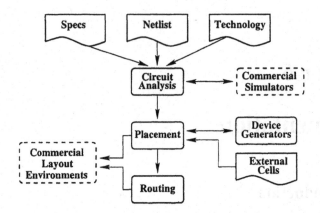

Figure 6.1: Software architecture of the analog layout tool LAYLA

- *module generators (36000 LOC)* : procedural generators for basic device layout structures

- *placement (26000 LOC)* : the placement algorithm as described in chapter 4. This includes the code to execute the various moves, the cost function and the simulated annealing engine.

- *router (21000 LOC)* : the routing algorithm as described in chapter 5.

The software architecture of the LAYLA tool is shown in Fig. 6.1. The LAYLA program is designed in an object oriented way. The use of the object oriented design style results in a system that is easy to maintain and to extend. All module generators for instance are derived from one base class. Additional module generators can be added to the system by deriving another class from the base class and implementing the function that generates the actual layout. None of the software that manipulates the modules has to be changed.

6.2.2 Interface to Electronic Design Frameworks

During the development of the LAYLA system, we have used several commercial layout environments to display and manipulate the results of our tools (see Fig. 6.1). Currently, we have an interface to the Mentor Graphics Falcon Framework [Falcon 97], the Cadence Design Framework II [DFII 97] and the Rockwell Object Symbolic Environment (ROSE) [ROSE 97]. For each framework integration, we have implemented a graphical user interface that allows to call the different layout tools from within the environment and an interface to load the results back into the framework. These routines are written in the extension language provided by the framework vendor. In the case of the Falcon Framework, the extension language is AMPLE [AMPLE 97] and for the Design Framework II, the extension language is SKILL [SKILL 97]. Both of these languages allow users to write extensions to their environments, using the framework's interface mechanisms. In the case of ROSE, the interface routines were written directly into the system.

LAYLA uses its own internal data structures to represent layouts. To import these data structures into commercial frameworks, we write the layout to a file as a set of extension language commands (AMPLE or SKILL). By executing the command file, we recreate the layout in the framework's layout environment.

6.3   Use of LAYLA in an Industrial Environment

An automatic layout tool has to be integrated into the full suite of tools that are used during production design. In this section we describe some of the links between a layout program and the rest of the design tools.

6.3.1   Link to Schematic Capture

In an industrial environment, a design is captured using a schematic entry tool. The design engineer constructs his design by placing and interconnecting symbols from a library and assigning values to the various properties that describe each symbol's electrical behavior. To provide a tight integration between electrical design and layout, the same schematic that is used for design and simulation must be used for layout as well. Therefore, it must be possible to annotate the layout constraints to the schematic. This can be done by extending the symbol libraries to include properties that are relevant for the layout tools. Examples of these properties are the various properties that drive the device generators, the properties that are used to define matching and symmetry groups and the properties that define port locations. For easy verification and inspection of the layout, a cross-probing mechanism between the layout and schematic must be provided.

6.3.2   Link to Simulation

With the ever decreasing size of integrated circuit devices, the influence of the parasitic elements associated with the layout of devices, such as series resistances and capacitances becomes more and more noticeable. For example, the *pn* junctions which form the MOS device source and drain regions each have a nonlinear voltage dependent capacitance that depends on the layout of the transistor. Passive elements such as resistors and capacitors have other parasitic components associated with them. The values of these parasitic components are layout dependent. For high-performance analog and RF designs, these layout dependent device parasitics have to be taken into account during electrical design. The use of procedural device generators allows to predict the device parasitics before the layout is done. To provide accurate simulation results, our device library has been synchronized with a library of simulation models. For each device generator in the library, a corresponding simulation model has been developed. This model includes the layout dependent parasitics. Examples of these composite models are shown in Fig. 6.2 for a capacitor and a resistor.

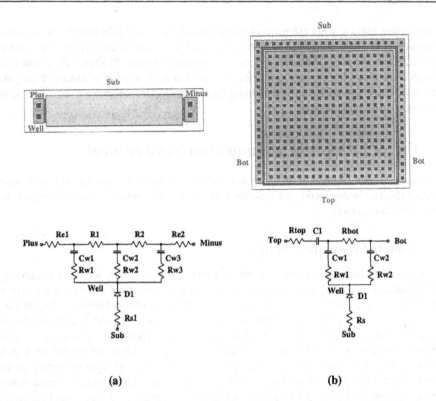

Figure 6.2: Layout models for (a) a resistor and (b) a capacitor.

6.3.3   Back-Annotation of Layout Parasitics

The use of our automatic layout system allows to generate the layout of a circuit in a couple of minutes. This makes it possible for a designer to quickly generate a layout of his circuit at various stages of the design process. The parasitics computed by the layout tools can be back annotated to the schematic to re-simulate the circuit and to evaluate the performance degradation of a particular layout. When used like this, an automatic layout tool tightens the link between electrical and physical design.

6.4   Results

Manual circuit level layout can be subdivided into module generation, placement, routing and verification. For a typical design, the layout engineer spends about 50% of his time in module generation, 12.5% in placement, 25% in routing and another 12.5% in verification. Through the use of a schematic driven layout methodology, combined with a library of device generators, the time spent in module generation can be reduced to virtually nothing. The use of module generators also results in layout that is correct by construction, which reduces the time spent in verification of the design. Interactive use of the place and route algorithms proposed in this book significantly reduces the time spent in placement and routing. The overall result of the introduction of LAYLA in the industrial environment was a 5 to 7x productivity gain compared with manual layout. This has been consistently observed during the layout of over 20 mixed-signal and RF chips.

Chapter 7

General Conclusions

This work has addressed the problem of automatic layout generation for analog integrated circuits. In chapter 2 we have described the major layout parasitic effects that influence the performance of analog circuits : interconnect parasitics, device mismatch and thermal effects. All of these effects have to be taken into account simultaneously during layout in order to keep the performance degradation within specified limits. We have proposed a direct performance driven layout strategy that, when applied to the different layout steps, guarantees a fully functional layout that respects all performance constraints. The major novelty of the method is that it drives the layout tools directly by the performance constraints, without an intermediate parasitic constraint generation step. During placement and routing, the performance characteristics of the circuit are evaluated using a linear approximation based on performance sensitivities. Using this approach, a complete and sensible trade-off between different layout alternatives can be made on the fly, and the resulting circuit layout can be guaranteed to be correct.

Apart from this general result, some more concrete results have been obtained in each of the analog layout generation subproblems : module generation, placement and routing.

In general, three different strategies can be applied to module generation for analog circuit layout. The first strategy uses a large library of procedural module generators that implement most of the analog layout knowledge. The second strategy limits the number of procedural module generators as much as possible and relies on the placement tool to assemble the more complicated substructures. A third approach is to construct a limited number of promising alternative sets of stacks in advance and to select the best one during placement.

We have demonstrated that neither of these strategies does the job in an optimal way. A number of substructures returns over and over again in analog circuits and can not be assembled by merging elementary devices. For these components, specific procedural module generators have to be written. These module generators have to be combined with merging during placement to achieve close to optimal layouts in a fast, reliable and predictable way. As described in chapter 3, to overcome the problem of technology dependent module generator libraries, we have proposed a technique that isolates technology dependence in technology parameter files and device definitions. The LAYLA module generator library has been used with more than 20 different technologies from 5 different foundries without modifications to the source code.

The second important component of an automatic layout system is the placement tool. The placement phase is crucial since all of the layout parasitics that determine the performance degradation of an analog circuit layout (interconnect parasitics, device mismatch and thermal effects) are either determined or greatly influenced by the placement of the circuit. Using our direct performance driven layout strategy, we have created the first analog placement tool that handles interconnect parasitics, device mismatch and thermal effects simultaneously and directly, without an intermediate constraint generation step, as described in chapter 2. In addition to that, our placement tool also supports other constraints that are crucial for high quality analog layout : symmetric placement, aspect-ratio and terminal position constraints, arbitrary rectangular component shapes, dynamic performance driven merging of diffusions and simultaneous location and shape optimization. The functionality of the tool has been demonstrated with several industrial example circuits.

After the placement phase, it is up to the router to add the actual wires to the layout and hence to fix the final values of the interconnect parasitics. The main requirement for the router is thus to limit the actual performance degradation within the specifications. It was shown in chapter 5 that routing also has a profound impact on the yield and testability of the circuit. The main result of our work on routing is a tool that controls the routing parasitics such that the overall performance degradation stays within the specifications, and at the same uses any remaining performance margin to optimize the manufacturability of the layout. Manufacturability optimization is done by routing nets such that the probability of occurrence of hard to detect faults is minimized. A new criterion to quantify the detectability of a fault was developed. Two analog circuits were routed using this technique, and it was shown that the yield and the testability of the circuit layout improved significantly, while the performance remained within the specifications.

The tools and algorithms described in this thesis have been integrated in the automatic analog layout system LAYLA, and have been used in an industrial environment for more than a year. During this period, the tools have been used for the layout of more than 20 mixed-signal and RF chips. A 5 to 7x productivity gain over manual layout has been observed.

Suggestions for Future Work

The interconnect parasitics that are taken into account in our layout tools are interconnect capacitance and resistance. Although this approximation is sufficiently accurate for low and medium frequency applications, for RF design, the inductance of the wires can no longer be neglected. The performance driven layout strategy introduced in this work is capable of handling any type of layout parasitic. To include parasitic wire inductance in the performance calculations, fast and accurate techniques to extract parasitic inductance have to be developed. The interconnect model presented in chapter 2 has to be extended to include the parasitic self inductance of each wire and the mutual inductance between every pair of circuits nodes. Although the inclusion of inductance effects would significantly increase the complexity of the performance driven layout problem, we believe that is possible to do so and that this extension would allow the use of automatic layout techniques in the growing domain of RF circuit design.

The layout tools presented in this work are capable of generating performance correct layouts for a broad range of applications. In general however, we noticed a 10 to 15 % area penalty when

comparing the results of our automatic layout tool with manual layout. Upon close examination of the layout, the main reason for this area penalty appears to be the artificial separation of the place and route phases. Experienced analog layout engineers do both tasks simultaneously : they place components with routing in mind and insert wires such that subsequent placement of components becomes easy. An important consequence of separating the placement and routing steps in the macro-cell layout style is that the placement algorithm is responsible for allocating the routing area. Failure to allocate sufficient interconnect area results in unroutable placements and therefore, the routing area is usually slightly overestimated, resulting in loss of density. Although the dynamic routing area estimation technique we presented in chapter 4 is a significant improvement over static approaches, we still believe that a lot can be gained in this area. One solution to the problem would be to perform placement and routing simultaneously. Early attempts in this direction [Cohn 94] revealed that the main drawback of this approach is the complexity of the problem, which limits its application domain to small circuits. Other approaches which can be investigated for larger circuits include combining a global routing algorithm within the placement optimization loop or the use of compaction after routing. The latter solution requires a sophisticated compaction algorithm that is capable of handling performance constraints and various other analog constraints such as symmetry and matching.

Integration of electrical and physical design is another grey area in analog synthesis. During electrical design, the effect of layout parasitics is usually approximated or discarded totally. For high-performance analog and RF design, designers usually include estimates of layout parasitics during electrical design. The estimated values of the layout parasitics are based on experience and on the estimated size of the circuit. Overestimation of layout parasitics results in wasted power and area, while underestimation of parasitics leads to physically un-realizable circuits. The automatic layout tools presented in this thesis could be used to derive more meaningful estimates of layout parasitics. The speed of the layout generation process is currently too slow to call in the inner loop of an automatic synthesis process. A template based or procedural approach could be explored for this purpose.

Bibliography

[AMPLE 97] AMPLE Reference Manual, Mentor Graphics, 1997

[Alb 95] J. Albers, "An Exact Recursion Relation Solution for the Steady-State Surface Temperature of a General Multilayer Structure," IEEE Trans. on Comp., Pack., and Manuf. Tech. - Part A, vol. 18, no. 1, pp. 31-38, March 1995.

[And 58] T. Anderson, "An Introduction to Multivariate Statistical Analysis," John Wisley and Sons, 1958.

[Arno 88] M. Arnold, W. Scott, "An Interactive Maze Router with Hints," Proc. 25th ACM/IEEE Design Automation Conf., pp. 672-676, 1988.

[Arora 96] N. D. Arora, K. V. Raol, L. M. Richardson, "Modeling and Extraction of Interconnect Capacitances for Multilayer VLSI Circuits," IEEE Trans. on Computer-Aided Design, vol. CAD-15, no. 1, Jan. 1996.

[Bak 90] H. Bakoglu, *Circuits, Interconnections and Packaging for VSLI* Addison Wesley, 1990.

[Bark 88] E. Barke, "Line-to-Ground Capacitance Calculation for VLSI: a Comparison," IEEE Trans. on Computer-Aided Design, Vol. CAD-7, No. 2, pp. 295-298, Feb. 1988.

[Bas 93] B. Basaran, R. Rutenbar, L. Carley, "Latchup-Aware Placement and Parasitic-Bounded Routing of Custom Analog Cells," Proc. IEEE Intl. Conf. on Computer-Aided Design, pp. 415-421, Nov. 1993.

[Bas 96] B. Basaran, R. Rutenbar, "An O(n) Algorithm for Transistor Stacking with Performance Constraints," Proc. 33rd ACM/IEEE Design Automation Conf., pp. 221-226, June 1996.

[Bas 96b] J. Bastos, M. Steyaert, B.Graindourze, W.Sansen, "Matching of MOS Transistors with Different Layout Styles," Proc. IEEE Int. Conference on Microelectronic Test Structures, pp. 17-18, March 1996.

[Bast 95] J. Bastos, M.Steyaert, R. Roovers, P.Kinget, W. Sansen, B.Graindourze, A.Pergoot, E. Janssens, "Mismatch Characterization of Small Size MOS Transistors," Proc. IEEE Int. Conference of Microelectronic Test Structures, Vol. 8, pp. 271-276, March 1995.

[Bast 96a] J. Bastos, M. Steyaert and W. Sansen, "A High Yield 12-bit 250-MS/s CMOS D/A Converter," Proc. IEEE Custom Integrated Circuits Conf, pp. 20.6.1-20.6.4, May 1996.

[Bast 96c] J. Bastos, M. Steyaert, B.Graindourze, W.Sansen, "Influence of Die Bonding on MOS Transistor Matching," Proc. IEEE Int. Conference on Microelectronic Test Structures, pp. 27-31, March 1996.

[Ben 76] P. Benedeck, "Capacitances of a Planar Multiconductor Configuration on a Dielectric Substrate by a Mixed Order Finite-Element Method," IEEE Trans. on Circuits and Systems, vol. CAS-23, no. 5, pp. 279-283, May 1976.

[Cath 88] F. Catthoor, H. de Man, J. Vandewalle, "SAMURAI: a general and efficient simulated-annealing schedule with fully adaptvie annealing parameters," Integration, the VLSI journal, vol. 6, pp. 147-178, 1988.

[Chak 90] S. Chakravarty, X. He, S. Ravi, "On Optimizing nMOS and Dynamic CMOS Functional Cells," Proc. IEEE Intl. Symp. on Circuits and Systems, pp. 1701-1704, May 1990.

[Chang 95] J. C. Chang, M. A. Styblinski, Yield and Variability Optimization of Integrated Circuits, Kluwer Academic Publishers, Boston, Dordrecht, London, February 1995.

[Charbon 92] E. Charbon, E. Malavasi, U. Choudhury, A. Casotto, A. Sangiovanni-Vincentelli, "A Constraint-Driven Placement Methodology for Analog Integrated Circuits,", proc. IEEE Custom Integrated Circuits Conf., pp. 28.2.1-4, May 1992.

[Charbon 93] E. Charbon, E. Malavasi, A. Sangiovanni-Vincentelli, "Generalized Constraint Generation for Analog Circuit Design," Proc. Intl. Conf. on Computer-Aided Design, pp. 408-414, November 1993.

[Charbon 94] E. Charbon, G. Holmlund, B. Donecker, A. Sangiovanni-Vincentelli, "A Performance-Driven Router for RF and Microwave Analog Circuit Design," Proc. IEEE ICCAD, Nov 1994.

[Charbon 94a] E. Charbon, E. Malavasi, D. Pandini, A. Sangiovanni-Vincentelli, "Imposing Tight Specifications on Analog IC's through Simultaneous Placement and Module Optimization," Proc. IEEE Custom Integrated Circuits Conf., pp. 525-528, May 1994.

[Charbon 94b] E. Charbon, E. Malavasi, D. Pandini, A. Sangiovanni-Vincentelli, "Simultaneous Placement and Module Optimization of Analog IC's," Proc. ACM/IEEE Design Automation Conf., pp. 31-35, June 1994.

[Chaw 70] B. Chawla, H. Gummel, "A Boundary Technique for Calculation of Distributed Resistance," IEEE Trans. on Electron Devices, vol. ED-17, pp. 915-925, Oct. 1970.

[Cheng 89] S. Cheng, P. Manos, "Effects of Operating Temperature on Electrical Parameters in an Analog Process," IEEE Circuits and Devices Magazine, pp. 31-37, July 1989.

[Choudhury 88] U. Choudhury, "Sensitivity Computation in SPICE3," Masters Thesis, U.C. Berkeley, Dec. 1988.

[Choudhury 90a] U. Choudhury, A. Sangiovanni-Vincentelli, "Use of Performance Sensitivities in Routing of Analog Circuits," proc. IEEE Intl. Symp. on Circuits and Systems, pp. 348-351, May 1990.

[Choudhury 90b] U. Choudhury and A. Sangiovanni-Vincentelli, "Constraint generation for routing analog circuits," Proc. ACM/IEEE Design Automation Conf., pp. 561-566, June 1990.

[Choudhury 90c] U. Choudhury and A. Sangiovanni-Vincentelli, "Constraint-based channel routing for analog and mixed analog/digital circuits," Proc. IEEE Intl. Conf. on Computer Aided Design, pp. 198-201, November 1990.

[Choudhury 91] U. Choudhury and A. Sangiovanni-Vincentelli, "An Analytical-Model Generator for Interconnect Capacitances," Proc. IEEE Custom Integrated Circuits Conf, pp. 8.6.1-8.6.4, May 1991.

[Choudhury 93] U. Choudhury, A. Sangiovanni-Vincentelli, "Automatic Generation of Parasitic Constraints for Performance-Constrained Physical Design of Analog circuits," IEEE Trans. on Computer-Aided Design, vol. 12,no. 2, pp. 208-224, February 1993.

[Choudhury 95] U. Choudhury, A. Sangiovanni-Vincentelli, "Automatic Generation of Analytical Models for Interconnect Capacitances," IEEE Trans. on Computer-Aided Design, vol. CAD-14, no. 4, April 1995.

[Cohn 91] J. M. Cohn, R. A. Rutenbar, and L. R. Carley, "KOAN/ANAGRAM II: New tools for device-level analog placement and routing," IEEE J. Solid-State Circuits, pp. 330-342, no. 3, Mar. 1991.

[Cohn 94] J. Cohn, J. Garrod, R. Rutenbar, R. Carley, *Analog Device-Level Layout Automation*, Kluwer Academic Publishers, Boston, Dordrecht, London, 1994.

[Conway 92] J. D. Conway, G. G. Schrooten, "An automatic layout generator for analog circuits," Proc. European Design Automation Conf., pp. 513-519, 1992.

[DFII 97] Design Framework II, Cadence Design Systems, Inc.,1997

[Dier 82] W. H. Dierking, J. D. Bastian, "VLSI Parasitic Capacitance Determination by Flux Tubes," IEEE Circuits and Systems Mag., pp 11-18, March 1982.

[Dir 69] S. Director, R. Rohrer, "The Generalized Adjoint Network and Network Sensitivities," IEEE Trans. on Circuit Theory, vol. CT-16, pp.318-322.

[Dona 90] W. Donath, et al., "Timing Driven Placement Using Complete Path Delays," Proc. ACM/IEEE Design Automation Conference, pp. 84-89, 1990.

[Donnay 94a] S. Donnay, K. Swings, G. Gielen, W.Sansen, W. Kruiskamp, D. Leenaerts, "A methodology for Analog Design Automation in Mixed-Signal ASICs," Proc. IEEE European Design and Test Conference, pp. 530-534, March 1994.

[Donnay 94b] S. Donnay, K. Swings, G. Gielen, W. Sansen, "A Methodology for Analog High-Level Synthesis," Proc. IEEE Custom Integrated Circuits Conf, pp. 373-376, May. 1994.

[Donz 91] L.-O. Donzelle et al.,, "A Constraint Based Approach to Automatic Design of Analog Cells," Proc. 28th AMC/IEEE Design Automation Conf., June 1991.

[Dun 84] A. Dunlop, V. Agrawal, D. Deutsch, M. Jukl, P. Kozak, M. Wiesel, "Chip Layout Optimization Using Critical Path Weighting," Proc. ACM/IEEE Design Automation Conference, pp. 32-39, 1984.

[Falcon 97] Falcon Framework Release Notes, Mentor Graphics, 1997.

[Felt 93] E. Felt, E. Malavasi, E. Charbon, T. Totaro, A. Sangiovanni-Vincentelli, "Performance-Driven Compaction for Analog Integrated Circuits," Proc. IEEE Custom Integrated Circuits Conf., pp.1731-1735, May 1993.

[Fidu 82] C. Fiduccia, R. Mattheyses, "A Linear-Time Heuristic for Improving Network Partitions," Proc. IEEE/ACM Design Automation Conference, pp. 175-181, June 1982.

[Fisher 87] J. Fisher, Koch R., "A Highly Linear CMOS Buffer Amplifier," IEEE J. of Solid State Circuits, vol. SC-22,no. 3, pp. 330-334, June 1987.

[Fredman 87] M. L. Fredman, R. E. Tarjan, "Fibonacci Heaps and Their Uses in Improved Network Optimization Algorithms," Journal of the ACM, vol. 34, pp. 596-615, 1987.

[Fuka 76] K. Fukahori, P. R. Gray, "Computer Simulation of Integrated Circuits in the Presence of Electrothermal Interaction," IEEE Journal of Solid State Circuits, vol. SC-11, no. 6, pp. 834-846, Dec. 1976.

[Gad 91a] G. Gad-El-Karim, R. S. Gyurcsik, "Use of Performance Sensitivities in Analog Cell Layout," proc. IEEE Intl. Symp. on Circuits and Systems, pp. 2008-2011, May 1991.

[Gad 91b] G. Gad-El-Karim, R. S. Gyurcsik, "Generation of Performance Sensitivities for Analog Cell Layout," proc. ACM/IEEE Design Automation Conf., pp. 500-505, June 1991.

[Gao 91] T. Gao, P. Vaidya, C. Liu, "A New Performance Driven Placement Algorithm," Proc. IEEE Intl. Conf. on Computer-Aided Design, pp. 44-47, 1991.

[Geer 93] B. Geeraerts, W. Van Peteghem, W. Sansen, "A BiMOS diode matrix for the characterization of static and transient thermal phenomena on silicon," Proc. SEMITHERM-IX, pp. 108-111, February 1993.

[Genderen 96] A. J. van Genderen, N. P. van der Mejs, T. Smedes, "Fast Computation of Substrate Resistances in Large Circuits," Proc. European Design and Test Conf., pp. 560-565, March 1996.

[Ghar 95a] R. Gharpurey, R. Meyer, "Modeling and Analysis of Substrate Coupling in Integrated Circuits," Proc. IEEE Custom Integrated Circuit Conf., pp. 7.3.1-7.3.4, May 1995.

[Ghar 95b] R. Gharpurey, R. Meyer, "Analysis and Simulation of Substrate Coupling in Integrated Circuits," Intl. Journ. of Circuit Theory and Applications, vol. 23, pp. 381-394, July-August 1995.

[Gielen 92] G. Gielen, K. Swings, W.Sansen, "Open Analog Synthesis System Based on Declarative Models," chapter 18 in *Analog Circuit Design, Operational Amplifiers, Analog to Digital Convertors, Analog Computer Aided Design*, Kluwer Academic Publishers, Boston, Dordrecht, London, 1992.

[Gielen 94] G. Gielen, Z. Wang, W. Sansen, "Fault Detection and Input Stimulus Determination for the Testing of Analog Integrated Circuits Based on Power-Supply Current Monitoring," Proc. IEEE ICCAD, pp. 495-498, November 1994.

[Gielen 95a] G. Gielen, G. Debyser, K. Lampaert, F. Leyn, K. Swings, G. Van der Plas, W. Sansen, D. Leenaerts, P. Veselinovic. W. van Bokhoven, "An Analog Module Generator for Mixed Analog/Digital ASIC design", Intl. Journal of Circuit Theory and Applications, pp. 269-283, July-August 1995.

[Gielen 95b] G. Gielen, G. Debyser, S. Donnay, K. Lampaert, F. Leyn, K. Swings, G. Van Der Plas, P. Wambacq, W. Sansen, "Comparison of Analog Synthesis using Symbolic Equations and Simulation," Proc. IEEE European Conf. on Circuit Theory and Design, 1995.

[Gielen 96] G. Gielen, F. Franca, "CAD Tools for Data Converter Design: An Overview," IEEE Trans. on Circuits and Systems-II: Analog and Digital Signal Processing, vol. 43, no. 2, pp. 77-89, Feb. 1996.

[Gray 71] P. R. Gray, D. J. Hamilton, "Analysis of electrothermal integrated circuits," IEEE Journal of Solid State Circuits, vol. SC-6, pp. 8-14, Feb. 1971.

[Gyur 90] R. Gyurcsik, S. Cochran, D.Thomas, "Performance-Driven Evaluation of Bipolar Analog Layouts," proc. IEEE Intl. Symp. on Circuits and Systems, pp. 827-829, May 1990.

[Had 75] F. Hadlock, "Finding a maximum cut of a planar graph in polynomial time," SIAM Journ. of Computing, vol. 4, no.3, pp. 221-225, September 1975.

[Hall 87] J. Hall, D. Hocevar, P. Yang, M. McGraw, "SPIDER - A CAD System for Modeling VLSI Metallization Patterns," IEEE Trans. on Computer Aided Design, vol. CAD-6, no. 6, pp. 1023-1031, Nov. 1987.

[Han 72] M. Hanan, J. Kurtzberg, *Design Automation of Digital Systems*. M. Breuer, Ed., Prentice Hall, Englewood Cliffs, N. J., Chap. 5, pp. 213-282. 1972.

[Harb 86] M. Harbour, J. Drake, "Calculation of Multiterminal Resistances in Integrated Circuits," IEEE Trans. on Circuits and Systems, vol. CAS-33,no. 4, pp. 462-465, April 1986.

[Hau 87] P. Hauge, R. Nair, E. Yoffa, "Circuit Placement for Predictable Performance," Proc. IEEE Intl. Conf. on Computer-Aided Design, pp. 88-91, 1987.

[Heyns 80] W. Heyns, W. Sansen, H. Beke, "A line-expansion algorithm for the general routing problem with a guaranteed solution," Proc. 17th ACM/IEEE Design Automation Conf., pp.143-249, 1980.

[High 69] D.W. Hightower, "A solution to line-routing problems on the continuous plane," Proc. 6th Design Automation Workshop, pp. 1-24, 1969.

[Hoc 85] D. Hocevar, P. Yang, T. Trick, B. Epler, "Transient Sensitivity Computation for MOSFET Circuits," IEEE Trans. on Electron Devices, vol. ED-32, no. 10, Oct. 1985.

[Hong 90] S. Hong, P. Allen, "Performance driven analog layout compiler," proc. IEEE Intl. Symp. on Circuits and Systems, pp. 835-838, May 1990.

[Hor 83] M. Horowitz, R. Dutton, "Resistance Extraction for Mask Layout Data," IEEE Trans. on Computer-Aided Design, vol. CAD-2, no. 3, pp. 145-150, July 1983.

[Host 85] B. J. Hosticka, K.-G. Dalsab, D. Krey, G. Zimmer, "Behavior of Analog MOS Integrated Circuits at High Temperatures," IEEE Journal of Solid State Circuits, vol. SC-20, no. 4, pp. 871-874, Aug. 1985.

[Jack 89] M. Jackson, E. Kuh, "Performance-Driven Placement of Cell Based IC's," Proc. ACM/IEEE Design Automation Conference, pp. 370-375, 1989.

[Jeps 84] D. W. Jepsen and C. D. Gelatt Jr, "Macro placement by Monte Carlo Annealing," Proc. IEEE int. Conf. on Computer Design, pp. 495-498, Nov. 1984.

[Joarder 94] "A Simple Approach to Modeling Cross-Talk in Integrated Circuits," IEEE Journ. of Solid-State Circuits, vol. SC-29, no. 10, pp. 1212-1219, Oct. 1994.

[Johnson 84] "Chip Substate Resistance Modeling Technique for Integrated Circuit Design", IEEE Trans. on Computer-Aided Design, vol. CAD-3, pp. 126-134, April 1984.

[Kayal 88] M. Kayal, S. Piguet, M. Declercq, B. Hochet, "SALIM: a layout generation tool for analog ICS," proc. CICC, pp. 7.5.1-4, 1988.

[Ker 70]   B. Kernighan, S. Lin, "An Efficient Heuristic Procedure for Partitioning Graphs," Bell Systems Technical Journal, Vol. 49, No. 2, pp. 291-308, 1970.

[King 96] P. Kinget, M. Steyaert, "Impact of transistor mismatch on the speed-accuracy-power trade-off of analog CMOS circuits," Proc. IEEE Custom Integrated Circuit Conf., pp. 333-336, May 1996.

[Kirk 83] S. Kirkpatrick, C. D. Gelatt, and M. P. Vecchi, 'Optimization by simulated annealing," Science, vol. 220, no. 4598, pp. 671-680, May 1983.

[Kokkas 74] A. Kokkas, "Thermal Analysis of Multiple-Layer Structures," IEEE Trans. on Electron Devices, vol. ED-21, no. 11, November 1974.

[Krus 56] J. Kruskal, "On the Shortest Spanning Subtree of a Graph and the Travelling Salesman Problem," Proc. Am. Math. Soc., vol. 7, pp. 48-50, 1956.

[Kuhn 87] J. Kuhn, "Analog module generators for silicon compilation," VLSI Systems design, May 1987.

[Kuo 93]   S.-Y. Kuo, "YOR: A Yield-Optimizing Routing Algorithm by Minimizing Critical Areas and Vias," IEEE Trans. Computer-Aided Design, vol. 12, NO. 9, September 1993.

[LAYLA man] K. Lampaert, LAYLA user manual, 1996.

[Laar 87] P. van Laarhoven, E. Aarts, Simulated Annealing: Theory and Applications, Kluwer Academic Publishers, Boston, Dordrecht, London, 1987.

[Laar 87] P. van Laarhoven, E. Aarts, Simulated Annealing: Theory and Applications, Kluwer Academic Publishers, 1987.

[Laker-Sansen 94] K. R. Laker, W. M. C. Sansen, Design of Analog Integrated Circuits and Systems, McGraw-Hill, Inc., New York, 1994.

[Lampaert 94] K. Lampaert, G. Gielen, W. Sansen, "Performance-Driven Placement of Analog Circuits," Proc. IEEE European Solid-State Circuits Conf., pp. 156-159, September 1994.

[Lampaert 95a] K. Lampaert, G. Gielen, W. Sansen, "Direct Performance-Driven Placement of Mismatch-Sensitive Analog Circuits," Proc. IEEE European Design and Test Conf., pp. 597, March 1995.

[Lampaert 95b] "Direct Performance-Driven Placement of Mismatch-Sensitive Analog Circuits," Proc. ACM/IEEE Design Automation Conf., pp. 445-449, June 1995.

[Lampaert 95c] K. Lampaert, G. Gielen , W. Sansen, "A Performance-Driven Placement Tool for Analog Integrated Circuits," IEEE Journ. of Solid-State Circuits, vol. SC-30, no. 7, pp. 773-780, July 1995.

[Lampaert 96a] K. Lampaert, G. Gielen, W. Sansen, "Analog Routing for Manufacturability," Proc. IEEE Custom Integrated Circuits Conf., May 1996.

[Lampaert 96b] K. Lampaert, G. Gielen, W. Sansen, "Thermally Constrained Placement of Analog and Smart Power Integrated Circuits," Proc. IEEE European Solid-State Circuits Conf., Sep. 1996.

[Lampaert 97a] K. Lampaert, G. Gielen, W. Sansen, "Thermally Constrained Placement of Smart-Power IC's and Multichip Modules," accepted for publication in Proc. SEMI-THERM, 1997.

[Lampaert 97b] K. Lampaert, G. Gielen, W. Sansen, "Analog Routing for Performance and Manufacturability," submitted to IEEE Journ. of Solid-State Circuits.

[Lee 61] C.Y. Lee, "An algorithm for path connections and its app.lication," IRE Trans. on electronic computers, vol. EC-10, pp. 346-365, 1961.

[Lee 88] C. Lee, A. Palisoc, "Real-Time Thermal Design of Integrated Circuit Devices," IEEE Trans. on Components, Hybrids, and Manufacturing Techn., vol. 11, no. 4, December 1988.

[Lee 89] C. Lee, A. Palisoc, J. Min, "Thermal Analysis of Integrated Circuit Devices and Packages," IEEE Trans. on Components, Hybrids and Manufacturing Techn., vol. 12, no. 4, December 1989.

[MS 92] HSpice User's Manual, Meta-Software, 1992.

[Malavasi 90] E. Malavasi, U. Choudhury and A. Sangiovanni-Vincentelli, "A routing methodology for analog integrated circuits," Proc. IEEE Intl. Conf. on Computer Aided Design, pp. 202-205, November 1990.

[Malavasi 91] E. Malavasi, E. Charbon, G. Jusuf, R. Totaro, A. Sangiovanni-Vincentelli, "Virtual Symmetry Axes for the Layout of Analog IC's," Proc. ICVC, pp. 195-198, Oct. 1991.

[Malavasi 93] E. Malavasi, A. Sangiovanni-Vincentelli, "Area Routing for Analog Layout," IEEE Trans. on Computer-Aided Design, vol. 12, no. 8, pp. 1186-1193, August 1993.

[Malavasi 95] E. Malavasi, D. Pandini, "Optimum CMOS Stack Generation with Analog Constraints," IEEE Trans. on Computer-Aided Design, vol. 14, no. 1, January 1995.

[Malavasi 95] E. Malavasi, A. Sangiovanni-Vincentelli, "Dynamic Bound Generation for Constraint-Driven Routing," Proc. IEEE CICC, pp. 477-480, May 1995.

[Maly 86] W. Maly, A. J. Strojwas, S. W. Director, "VLSI Yield Prediction and Estimation: A Unified Framework," IEEE Trans. on Computer-Aided Design, pp. 114-130, Jan. 1986.

[Maly 86] W. Maly, A. J. Strojwas, S. W. Director, "VLSI Yield Prediction and Estimation : A Unified Framework," IEEE Trans. Computer-Aided Design, vol. CAD-5, no. 1, pp. 114-130, Jan. 1986.

[Maly 90] W. Maly, "Computer-Aided Design for VLSI Circuit Manufacturability," Proc. of the IEEE, vol. 78, no. 2, pp. 356-391, Feb. 1990.

[Mar 89] M. Marek-Sadowska, S. Lin, "Timing Driven Placement," Proc. IEEE Intl. Conf. on Computer-Aided Design," pp. 94-97, 1989.

[Mar 90] D. Marple, M. Smulders, H. Hegen, "Tailor: A Layout System Based on Trapezoidal Corner Stitching," IEEE Trans. on Computer-Aided Design, vol. 8, no. 1, pp. 66-90, January 1990.

[Marg 87] A. Margarino, A. Romano, A. de Gloria, F. Curatelli, P. Antognetti, "A Tile-Expansion Router," IEEE Trans. on Computer-Aided Design, vol. CAD-6, no. 4, pp. 507-517, July 1987.

[McNu 94] M. McNutt, S. LeMarquis, J. Dunkley, "Systematic Capacitance Matching Erros and Corrective Layout Procedures," IEEE J. of Solid State Circuits, vol. 29, no. 5, pp. 611-616, May 1994.

[Met 53] N. Metropolis, A. Rosenbluth, M. Rosenbluth, A. Teller, E. Teller, "Equation of State Calculation by Fast Computing Machines," Journ. of Chem. Physics, vol. 21, pp. 1087-1092, 1953.

[Meyer 93] V. Meyer zu Bexten, C. Morage, R. Klinke, W. Brockherde, K. Hess, "ALSYN : Flexible Rule-Based Layout Synthesis for Analog IC's" IEEE J. Solid-State Circuits, vol. 28, no. 3, pp. 261-268, March 1993.

[Mich 92] C. Michael, M. Ismail, "Statistical modeling of device mismatch for analog MOS integrated circuits," IEEE J. of Solid State Circuits, vol. SC-27, no. 2, pp. 154-165, February 1992.

[Mika 68] K. Mikami, K. Tabuchi, "A computer program for optimal routing of printed circuit connectors," Proc. IFIPS, vol. H47, pp. 1475-1478, 1968.

[Milor 89] L. Milor, V. Visvanathan, "Detection of Catastrophic Faults in Analog Integrated Circuits," IEEE Trans. on Computer-Aided Design, vol. CAD-8, no. 2, Feb. 1989.

[Mogaki 89] M. Mogaki, N. Kato, Y. Chikami, N. Yamada, Y. Kobayashi, "LADIES: An Automatic Layout System for Analog LSI's," Proc. IEEE Intl. Conf. on Computer-Aided Design, pp. 450-453, November 1989.

[Nab 91]   K. Nabors, J. White, "FastCap: A Multipole Accelerated 3-D Capacitance Extraction Program," IEEE Trans. on Computer-Aided Design, vol. CAD-10, no. 11, Nov. 1991.

[Nab 92]   K. Nabors, S. Kim, J. White, "Fast Capacitance Extraction of General Three-Dimensional Structures," IEEE Trans. on Microwave Theory and Techniques, vol. MTT-40, no. 7, July 1992.

[Nair 89] R. Nair, L. Berman, P. Hauge, E. Yoffa, "Generation of Performance Constraints for Layout," IEEE Trans. on Computer-Aided Design, vol. 8, no. 8, Aug. 1989.

[Nils 71] N. J. Nilsson, Problem-solving Methods in Artificial Intelligence, McGraw-Hill, Ch3, pp. 43-78, 1971.

[Ning 87] Q. Ning, P. M. Dewilde, F. L. Neerhoff, "Capacitance Coefficients for VLSI Multi-level Metallization Lines," IEEE Trans. on Electron Devices, vol. ED-34, pp. 644-649, 1987.

[Ogaw 86] Y. Ogawa, T. Ishii, Y. Shiraishi, H. Terai, T. Kozawa, K. Yuyama, K. Chiba, "Efficient Placement Algorithms Optimizing Delay For High-Speed ECL Masterslice LSI's," Proc. ACM/IEEE Design Automation Conference, pp. 404-410, 1986.

[Oht 86]   T. Ohtsuki (editor), Layout Design and Verification, North Holland, 1986.

[Otten 84] R. Otten, L. van Ginneken, "Floorplan Design using Simulated Annealing," Proc. IEEE Intl. Conf. on Computer-Aided Design, pp. 96-98, November 1984.

[Oust 84] J. Ousterhout, "Corner stitching: a data-structuring technique for VLSI layout tools," IEEE Trans. Computer-Aided Design, vol. CAD-3,no. 1, pp. 87-100, January 1984.

[Pap 91]   A. Papoulis, Probability, Random Variables and Stochastic Processes (third edition), McGraw-Hill International Editions, 1991.

[Peeters 93] E. Peeters, Ghafoor K., "Design of a Fully Differential High-Speed CMOS Amplifier," KU Leuven Masters Thesis. June 1993.

[Pel 89]   M. J. M. Pelgrom, A. C. J. Duinmaijer, A. P. G. Welbers, "Matching properties of MOS transistors," IEEE J. of Solid State Circuits, vol. SC-24,no. 5, pp. 1433-1440, October 1989.

[Pit 89]   A. Pitaksannonkul, S. Thanawastien, C. Lusinsap, J. A. Gandhi, "DTR: A Defect-Tolerant Routing Algorithm," Proc. 26th ACM/IEEE Design Automation Conf., pp. 795-798, June 1989.

[Prieto 97] J. A. Prieto, A. Rueda, J. M. Quintana, J. L. Huertas, "A Performance-Driven Placement Algorithm with Simultaneous Place and Route Optimization for Analog IC's" Proc. IEEE European Design and Test Conference, March 1997.

[Prim 57] R. Prim, "Shortest Connection Networks and Some Generalizations," Bell Systems Technical Journ., vol. 36, pp. 1389-1401, 1957.

[ROSE 97] Rockwell Object Symbolic Environment, Reference Manual Rockwell Semiconductor Systems, June 1997.

[Rao 86] V. Rao, A. Djordjevic, T. Sarkar, Naiheng, "Analysis of Arbitrarily Shaped Dielectric Media over a Finite Ground Plane," IEEE Trans. on Microwave Theory and Techniques, vol. MTT-33, pp. 472-475, 1986.

[Rijm 88] J. Rijmenants, T. Schwartz, J. Litsios, R. Zinszner, "ILAC: An automated layout tool for analog CMOS circuits," proc. Custom Integrated Circuits Conf., pp. 7.6.1-7.6.4, May 1988.

[Rijm 89] J. Rijmenants et al, "ILAC: An automated layout tool for analog CMOS circuits," IEEE J. Solid-State Circuits, pp. 417-425, no. 2, Apr. 1989.

[Rose 93] J. Rose, A. E. Gamal, A. Sangiovanni-Vincentelli, "Architecture of Field-Programmable Gate Arrays," Proceedings of the IEEE, vol. 81, no. 7, pp. 1013-1029, July 1993.

[Rueh 73] A.E. Ruehli and P.A. Brennan, "Efficient Capacitance Calculations for Three-Dimensional Multiconductor Systems," IEEE Trans. Microwave Theory Tech. MTT, Feb. 1973.

[SKILL 97] SKILL Language and Development, Cadence Design Systems, Inc.,1997

[Sahn 80] S. Sahni, A. Bhatt, "The Complexity of Design Automation Problems," Proc. IEEE/ACM Design Automation Conf., pp. 402-411, June 1980.

[Sato 87] M. Sato, "A Fast Line-Search Method Based on a Tile-Plane," Proc. IEEE Intl. Symp. on Circuits and Systems, pp. 588-591, 1987.

[Seid 88] A. Seidl, M. Svoboda, J. Oberndorfer, W. Rosner, "CAPCAL - A 3-D Capacitance Solver for Support of CAD Systems," IEEE Trans. on Computer-Aided Design, vol. CAD-7, no. 5, pp. 644-649, March 1987.

[Seq 93] C. Sequin, H. da Silva Facanha, "Corner-Stitched Tiles With Curved Coundaries," IEEE Trans. on Computer-Aided Design, vol. 9, no. 1, pp. 47-58, January 1993.

[Shah 91] K. Shahookar and P. Mazumder, "VLSI Cell Placement Techniques," ACM Computing Surveys, Vol. 23, No. 2, June 1991.

[Sher 95] N. Sherwani, *Algorithms for VLSI Physical Design Automation*, second edition, chapter 3, Kluwer Academic Publishers, Boston, Dordrecht, London, 1995.

[Shyu 84] J. B. Shyu, G. Temes, F. Krummenacher, "Random error effects in matched MOS capacitors and current sources," IEEE J. of Solid State Circuits, vol. SC-19,no. 6, pp. 948-955, December 1984.

[Shyu 86] J. B. Shyu, G. Temes, F. Krummenacher, "Characterization and modeling of mismatch in MOS transistors for precision analog design," IEEE J. of Solid State Circuits, vol. SC-21,no. 6, pp. 1057-1066, December 1986.

[Smedes 93a] "Substrate Resistance Extraction for Physics-Based Layout Verification," Proc. IEEE/PRORISC Workshop on CSSP, pp. 101-106, March 1993.

[Smedes 93b] T. Smedes, N. P. van der Mejs, A. J. van Genderen, "Boundary Element Methods for Capacitance and Substrate Resistance Calculations in a VLSI Layout Verification Package", *Software Applications in Electrical Engineering*, Ed.: P.P. Silvester, pp. 337-344, 1993.

[Smedes 95] T. Smedes, N. P. van der Mejs, A. J. van Genderen, "Extraction of Circuit Models for Substrate Cross-talk," Proc. IEEE Intl. Conf. on Computer-Aided Design, pp. 199-206, Nov. 1995.

[Sol 74] J. E. Solomon, "The Monolithic Op Amp: A Tutorial Study," IEEE Journal of Solid State Circuits, vol. SC-9, no. 6, pp. 314-332, Dec. 1974.

[Souk 78] J. Soukup, "Fast Maze Router," Proc. IEEE/ACM Design Automation Conference, 1984

[Stanisic 94] B. R. Stanisic, N. K. Verghese, D. J. Allstot, R. A. Rutenbar, L. R. Carley, "Addressing Substrate Coupling in Mixed-Mode ICs : Simulation and Power Distribution Synthesis," IEEE Journ. of Solid State Circuits, vol. SC-29,no. 3, pp. 226-237, April 1994.

[Stapper 83a] C. H. Stapper, "Modeling of Integrated Circuit Defect Sensitivities," IBM J. Res. Develop., vol. 27, no. 6, pp. 549-557, Nov. 1983.

[Stapper 83b] C. H. Stapper, F. M. Armstrong, K. Saji, "Integrated Circuit Yield Statistics," Proc. of the IEEE, vol. 71, no. 4, April 1983.

[Stapper 84] C. H. Stapper, "Modeling of Defects in Integrated Circuit Photolithographic patterns," IBM J. Res. Develop., vol. 28, no. 4, pp. 461-475, July 1984.

[Steyaert 93] M. Steyaert, R. Roovers and J. Craninckx, "A 110 MHz 8 bit CMOS Interpolating A/D converter," Proc IEEE Custom Integrated Circ. Conf., pp. 28.1.1-28.1.4 May 1993.

[Su 93] "Experimental Results and Modeling Techniques for Substrate Noise in Mixed-Signal Integrated Circuits," IEEE Journ. of Solid State Circuits, vol. SC-28,no. 4, pp. 420-430, April 1993.

[THOM 96] "POLYMIX TSKM Reference Manual," Thomson-CSF semiconducteurs specifiques.

[Tayl 85] C. D. Taylor, G. N. Elkhouri, T. E. Wade, "On the Parasitic Capacitances of Multilevel Parallel Metallization Lines," IEEE Trans. Electron Devices, vol. ED-32, no. 11, pp. 2408-2414, Nov. 1985.

[Thul 92] K. Thulasiraman, M. Swamy, *Graphs: theory and algorithms*, Wiley-Interscience, 1992.

[Ueb 86] R. H. Uebbing, M. Fukuma, "Process-Based Three-Dimensional Capacitance Simulation - TRICEPS," IEEE Trans. on Computer-Aided Design, vol. CAD-5, no. 1, pp. 215-220, Jan. 1986.

[Uehara 81] T. Uehara, W. van Cleemput, "Optimal Layout of CMOS Functional Arrays," IEEE Trans. on Computers, vol. C-30, no. 5, May 1981.

[Van Pet 93] W. Van Petegem, "Iterative Solutions to Electronic Problems, Based on Finite-Element Methods : Electro-Thermal Simulation and Electrical Impedance Tomography," PhD Thesis, Katholieke Universiteit Leuven, September 1993.

[Verghese 93] N. K. Verghese, D. J. Allstot, "Rapid simulation of substrate coupling effects in mixed-mode IC's," Proc. IEEE Custom Integrated Circuits Conf., pp. 18.3.1-18.3.4, May 1993.

[Verghese 95] "Fast Parasitic Extraction for Substrate Coupling in Mixed-Signal ICs," Proc. IEEE Custom Integrated Circuit Conf., pp. 121-124, May 1995.

[Vitt 85] E. Vittoz, "The Design of High-performance Analog Circuits on Digital CMOS Chips," IEEE Journal of Solid State Circuits, vol. SC 20, pp. 657-665, June 1985.

[Vlach 83] J. Vlach, K. Singhal, *Computer methods for circuit analysis and design.* Van Nostrand Reinhold, 1983.

[Wimer 87] S. Wimer, R. Pinter, J. Feldman, "Optimal Chaining of CMOS Transistors in a Functional Cell," IEEE Trans. on Computer-Aided Design, vol. CAD-6, No. 5, September 1987.

[Wong 86] D. Wong, C. Liu, "A new algorithm for floorplan design," Proc. IEEE/ACM Design Automation Conf., 1986.

[Wu 90] San-Yuan Wu, Sartah Sahni, "Covering Rectilinear Polygons by Rectangles," IEEE Trans. on Computer-Aided Design, vol. CAD-9, no. 4, pp. 377-388, April 1990.